Demokritos Scientific Journal
Open Access Research for Discovery

Special Edition

Featuring:

Exploring Quantum Entanglement through String Theory: Proposing Alternatives to Photon-Based Experiments

By Antonios Valamontes

Published by

Valamontes Center for Scientific Publication
Volume 1, Special Edition (2024)

Library of Congress Control Number (LCCN):

This journal has been assigned a Preassigned Control Number (PCN) by the Library of Congress. The PCN Program allows the Library to assign control numbers in advance of publication for titles that may be added to the Library's collections, ensuring that this work is properly cataloged and accessible for future research.

A Library of Congress control number (LCCN) is a unique identification number that the Library of Congress assigns to the catalog record created for each book in its cataloged collections. Librarians use it to locate a specific Library of Congress catalog record in the national databases.

Library of Congress Control Number: 2024920871
ISBN: 979 8340779533

© 2024 Valamontes Center for Scientific Publication. All rights reserved for journal design and presentation. The research work is open access and may be freely distributed.

Even if you are not a U.S. citizen and are not eligible to participate in the Library of Congress PCN Program, **Demokritos Scientific Journal** still offers significant advantages for publishing your work:

1. **Global Accessibility**: As an open-access platform, *Demokritos Scientific Journal* ensures that your research is accessible to a worldwide audience. This makes it an excellent option for researchers from any country who want to share their work without barriers.

2. **Open-Review Model:** The journal's unique open-review process allows for broader community engagement and feedback, enabling your work to gain visibility and scrutiny from the global scientific community, regardless of geographic boundaries.

3. **Prestige and Recognition**: Named after the renowned Greek philosopher **Demokritos**, the journal upholds a commitment to advancing the frontiers of scientific discovery. Publishing in a journal with such a legacy offers a prestigious platform for showcasing innovative research.

4. **Valamontes Center for Scientific Publication**: This independent publishing entity supports research from all regions, ensuring that your work is disseminated without being restricted by nationality or location. You can still be part of a respected and professional journal dedicated to advancing global physics research.

5. **Unique Platform for Specialized Research:** The journal caters to a niche audience that appreciates cutting-edge theoretical and applied physics, providing a focused platform for your work to reach its intended audience.

While participation in the Library of Congress PCN Program may be limited to U.S. authors and publishers, the benefits of publishing with *Demokritos Scientific Journal* extend far beyond national borders, giving your research the global reach and recognition it deserves.

Journal Description

This special edition of the *Demokritos Scientific Journal* presents groundbreaking research in the field of quantum entanglement through string theory, with a focus on proposing alternatives to traditional photon-based experiments. Authored by **Antonios Valamontes**, this paper explores innovative theoretical approaches to one of the most fascinating phenomena in modern physics, offering fresh insights into the behavior of entangled particles and their interactions within the framework of string theory.

As part of the journal's commitment to open-access research, this publication invites readers to engage with cutting-edge scientific discoveries. By purchasing this journal, you are supporting a platform dedicated to making advanced research accessible to the global scientific community.

Whether you are a physicist, researcher, or enthusiast, this edition serves as a valuable contribution to the ongoing discourse in theoretical physics, pushing the boundaries of what is known about quantum mechanics and beyond.

Table of Contents

Experimental Proposals **75**

    Particle Generation and Synchronization 75

        A1. Particle Accelerators: 75

        B1. Quantum Simulation: 78

        C1. Temporal Synchronization: 86

    Measurement and Correlation Analysis 92

        A2. Quantum Sensors: 92

        B2. Correlation Studies: 98

    Advanced Experimental Setups 105

        A3. Quantum Dot Arrays: 105

        A4. Superconducting Qubits: 110

        A5. High-Energy Physics Experiments: 116

    Potential Outcomes and Implications 122

        B1. Positive Results: 122

        B2. Negative Results: 127

    Interdisciplinary Approaches 132

        Development of Novel Quantum Algorithms 135

    References 139

Exploring Quantum Entanglement through String Theory: Proposing Alternatives to Photon-Based Experiments

By Antonios Valamontes

May 28, 2024

tony@valamontes.com

 Accessibility Notice for Scientific Papers: This logo indicates that the paper has been specifically formatted and written to accommodate readers with dyslexia and autism spectrum disorders (ASD). It ensures that the content is accessible and easier to read and understand for individuals with these conditions.[55]

Abstract

The primary motivation for exploring alternative hypotheses about quantum entanglement, particularly involving string theory characteristics, arises from the potential to overcome limitations and answer unresolved questions in existing photon entanglement experiments. This research proposes a significant expansion of our understanding of quantum entanglement by investigating the role of string theory in defining entanglement through shared characteristics such as vibrational modes, temporal synchronization, string tension, topological properties, dimensional embedding, gauge symmetries, and coupling constants.

Our hypothesis posits that particles achieve quantum entanglement if they share two or more of these fundamental characteristics. Positive results from high-energy physics experiments demonstrating significant correlations between particles with shared string properties would provide strong support for this hypothesis, suggesting a broader basis for quantum entanglement. Conversely, negative results would indicate the need to refine the hypothesis or explore additional factors influencing entanglement, thereby guiding future research in this field.

We will explore advanced experimental setups, including particle accelerators and quantum sensors, and develop novel quantum algorithms to analyze experimental data. Furthermore, interdisciplinary approaches involving condensed matter physics, quantum information science, and materials science will be leveraged to gain new perspectives and enhance the robustness of our experimental designs. This research aims to provide deeper insights into the fundamental nature of quantum entanglement and uncover new mechanisms governing entanglement across a wider range of particles and conditions.

Introduction

Quantum entanglement is a cornerstone of quantum mechanics, demonstrating non-local correlations between particles. Traditional experiments involving photon entanglement have shown robust violations of Bell's inequalities, confirming the presence of entanglement. However, these experiments face significant limitations, including decoherence, measurement issues, and potential biases introduced by specific experimental setups. To address these limitations, we propose a novel hypothesis that incorporates string theory characteristics to define entanglement.

Experimental Limitations

Photon entanglement experiments, such as those using spontaneous parametric down-conversion (SPDC), have demonstrated entanglement but are limited by specific conditions and setups. Decoherence, environmental interference, detector efficiency, and dark counts are some of the challenges that obscure the true nature of entanglement in these experiments. These limitations necessitate the exploration of alternative approaches to study entanglement comprehensively.

Theoretical Considerations

String theory, with its one-dimensional 'strings' vibrating at specific frequencies, not only provides a broader framework for understanding fundamental interactions but also offers valuable insights into the nature of entanglement. It suggests additional degrees of freedom and characteristics, such as vibrational modes and gauge symmetries, that could be crucial for our understanding of entanglement. This underscores the relevance of string theory in the context of our research.

Innovative Perspectives

Our research, which expands the scope of entanglement to include characteristics like vibrational modes, string tension, and gauge symmetries, not only offers new insights into quantum mechanics but also has practical implications. By addressing potential flaws in photon experiments, we can improve the robustness of our experimental designs. The interdisciplinary approaches and the development of novel quantum algorithms are key to this improvement.

Detailed Hypothesis

We hypothesize that particles achieve quantum entanglement if they share two or more fundamental characteristics derived from string theory. These characteristics include:

1. **Vibrational Modes:** Strings vibrate at specific frequencies, determining the properties of particles. Shared vibrational modes suggest a common quantum state, potentially leading to entanglement.
2. **Temporal Synchronization:** Alignment of particles' quantum states in time. Shared temporal states may lead to entanglement.
3. **String Tension:** The tension in a string affects its vibrational energy. Identical string tensions may facilitate entanglement due to shared energetic properties.
4. **Topological Properties:** Open or closed configurations of strings affect their interactions. Similar topological configurations may result in entanglement.
5. Dimensional Embedding: Strings exist in higher-dimensional spaces. Shared dimensional embeddings can be a basis for entanglement.
6. Gauge Symmetries: Symmetries governing strings' interactions. Shared gauge symmetries may lead to correlated behaviors indicative of entanglement.

7. **Coupling Constants:** The strength of interactions between strings is determined by coupling constants. Similar coupling constants may result in correlated quantum states.

Experimental Proposals

To test this hypothesis, we propose the following experimental approaches:

1. **Particle Generation and Synchronization:** Use high-energy particle accelerators to generate particles with specific string characteristics and synchronize them using lasers or atomic clocks.
2. **Measurement Techniques:** Employ quantum state tomography, spectroscopic analysis, and interferometry to measure the coupling constants and resulting quantum states.
3. **Quantum Sensors:** Utilize advanced quantum sensors and detectors to measure the properties of particles and analyze correlations.
4. **Quantum Dot Arrays and Superconducting Qubits:** Simulate strings with specific properties using quantum dot arrays and superconducting qubits to study entanglement.

Potential Outcomes and Implications

Positive Results: Demonstrating significant correlations between particles with shared string characteristics would support the hypothesis, suggesting a broader basis for quantum entanglement.

Negative Results: Lack of significant correlations would indicate the need to refine the hypothesis or explore additional factors influencing entanglement.

By exploring quantum entanglement through the lens of string theory, we aim to uncover new mechanisms and principles governing entanglement across a wider range of particles and conditions. This research will deepen our understanding of quantum mechanics and support the development of practical quantum technologies.

1. Experimental Limitations:

Photon Entanglement

Experiments involving photon entanglement, such as those using spontaneous parametric down-conversion (SPDC), have robustly demonstrated violations of Bell's inequalities, confirming the presence of entanglement. However, several experimental limitations raise questions about the completeness and universality of these results. These limitations often stem from the specific conditions and setups required for these experiments, which may not fully generalize to all quantum systems.

Specific Conditions and Setups

1.1. Spontaneous Parametric Down-Conversion (SPDC):

- Process: In SPDC, a single photon of higher energy (usually from a laser) is converted into two entangled photons of lower energy when it passes through a nonlinear crystal.
- Equation: The energy and momentum conservation in SPDC can be expressed as

$$E_p = E_s + E_i \quad \text{and} \quad \mathbf{k}_p = \mathbf{k}_s + \mathbf{k}_i$$

[1.0]

where E_p and $\mathbf{k}_p$ are the energy and momentum of the pump photon, and E_s, E_i and $\mathbf{k}_s, \mathbf{k}_i$ are the energies and momenta of the signal and idler photons, respectively.

1.2. Bell's Inequality and Measurement:

- Bell's Inequality: The violation of Bell's inequality is a key test for quantum entanglement. The CHSH form of Bell's inequality is given by

$$S = |E(\hat{a}, \hat{b}) - E(\hat{a}, \hat{b}') + E(\hat{a}', \hat{b}) + E(\hat{a}', \hat{b}')| \leq 2$$

[2.0]

where $E(\hat{a}, \hat{b})$ represents the correlation function between measurements along directions $\hat{a}$ and $\hat{b}$.

- Experimental Setup: Typical setups involve precise alignment of optical components, including nonlinear crystals, beam splitters, and photon detectors, all of which must be carefully calibrated to ensure accurate measurement of entanglement.

Limitations

1.2.1. Decoherence and Environmental Interference:
- Decoherence: Photons can interact with the environment, leading to decoherence, which destroys the entanglement. This interaction can be with air molecules, optical components, or even thermal radiation.
- Interference: External electromagnetic fields and vibrations can affect the stability of the experimental setup, leading to noise and measurement errors.

1.2.2. Detector Efficiency and Dark Counts:
- Efficiency: Photon detectors have limited efficiency, meaning not all entangled photons are detected. This inefficiency can lead to false negatives in entanglement detection.
- Dark Counts: Detectors also register false positives due to dark counts (spurious signals), which can be mistaken for genuine photon detections.

1.2.3. Loopholes in Bell Test Experiments:
- Locality Loophole: Ensuring that the choice of measurement settings and the actual measurements are space-like and separated (no causal connection) is challenging and crucial to closing the locality loophole.
- Detection Loophole: The detection efficiency must be high enough to ensure that the detected sample accurately represents the entire ensemble of entangled photons.

Mathematical Formalism and Completeness

1.3.1. Wavefunction and State Vector Representation:

The quantum state of entangled photons can be described by a wave function ψ or state vector |ψ⟩. For Case, the state of two entangled photons can be represented as

$$|\psi\rangle = \frac{1}{\sqrt{2}}(|H\rangle_1|H\rangle_2 + |V\rangle_1|V\rangle_2)$$

[3.0]

where H and V represent horizontal and vertical polarization states.

1.3.2. Density Matrix and Mixed States:
- Realistic scenarios often involve mixed states due to decoherence. The density matrix ρ provides a complete description of such states,

$$\rho = \sum_i p_i |\psi_i\rangle\langle\psi_i|$$

[4.0]

where p_i are probabilities and $|\psi_i\rangle$ are pure states.

Generalization to Other Quantum Systems

1.4.1. Limited Scope:
- While photon-based experiments have provided significant insights into quantum entanglement, their reliance on specific optical setups limits their generalizability to other quantum systems like electrons, atoms, or larger macroscopic systems.

1.4.2. Different Degrees of Freedom:
- Other quantum systems may have additional degrees of freedom (e.g., spin, orbital angular momentum) that are not easily accessible or measurable in photon-based systems. This complexity requires new theoretical and experimental approaches to study entanglement comprehensively.

While photon entanglement experiments have been instrumental in demonstrating the fundamental principles of quantum mechanics, they come with limitations that may not fully generalize to all quantum systems. Understanding these limitations and exploring alternative approaches, such as those inspired by string theory, can provide a more comprehensive view of entanglement and its underlying mechanisms.

Decoherence and Measurement Issues

Photon-based experiments are a cornerstone of quantum entanglement research. However, they face several challenges related to decoherence and measurement issues that can obscure the true nature of entanglement. These issues include environmental interactions that disrupt quantum states and the inherent limitations and potential biases in the measurement process.

Decoherence

Definition and Impact:
- Decoherence: Decoherence occurs when a quantum system interacts with its environment, causing the system to lose its quantum coherence. This process effectively transforms a pure quantum state into a mixed state, where the superposition of states is lost, and classical probabilities emerge.

- Impact on Entanglement: Decoherence can degrade or destroy entanglement between particles. For instance, entangled photons passing through a medium can interact with air molecules, dust particles, or thermal radiation, causing decoherence.

Mathematical Formalism:

- Density Matrix Representation: A pure quantum state $|\psi\rangle$ can be represented by a density matrix $\rho = |\psi\rangle\langle\psi|$. When decoherence occurs, the density matrix evolves into a mixed state

$$\rho = \sum_i p_i |\psi_i\rangle\langle\psi_i|$$

[5.0]

where p_i are the probabilities of different pure states $|\psi_i\rangle$.

- Master Equation: The time evolution of the density matrix under decoherence can be described by the Lindblad master equation

$$\frac{d\rho}{dt} = -\frac{i}{\hbar}[H, \rho] + \sum_k \left(L_k \rho L_k^\dagger - \frac{1}{2}\{L_k^\dagger L_k, \rho\} \right)$$

[6.0]

where H is the Hamiltonian of the system, and L_k are the Lindblad operators representing the interaction with the environment.

Cases in Photon Experiments:

- Polarization Decoherence: Photons entangled in polarization can experience decoherence due to birefringent materials in the optical path, which can unpredictably alter their polarization states.

- Path Decoherence: In interferometric setups, small vibrations or temperature fluctuations can cause path-length differences, leading to loss of coherence.

Measurement Issues

Measurement Process and Its Limitations:
- Photon Detection: Photon detection efficiency is crucial in entanglement experiments. Imperfect detectors with limited efficiency can miss detecting some entangled photons, leading to an incomplete dataset.
- Dark Counts: Photon detectors can register false counts due to thermal noise or cosmic rays, known as dark counts. These false positives can skew the results, making it seem as though there is entanglement where there might be none.

Mathematical Representation:
- Detection Efficiency: Let η represent the detection efficiency. The probability P_{det} of detecting a photon is

$$P_{det} = \eta P_{photon}$$

[8.0]

where P_{photon} is the probability of the photon's presence.

- Dark Counts: The observed count rate C_{obs} includes both genuine photon counts C_{photon} and dark counts C_{dark}

$$C_{obs} = C_{photon} + C_{dark}$$

[9.0]

Biases in Measurement:

- Detector Bias: Variations in detector efficiency across different detectors can introduce systematic biases. Ensuring uniform calibration and accounting for efficiency differences is essential to avoid skewed data.
- Timing Jitter: The uncertainty in the timing of photon detection, known as timing jitter, can affect the measurement of entangled states. This is particularly relevant in time-bin entanglement experiments.

Techniques to Mitigate Issues:

- High-Efficiency Detectors: Using detectors with higher efficiency and lower dark count rates can reduce measurement errors. Superconducting nanowire single-photon detectors (SNSPDs) are an Case of such advanced technology.
- Environmental Isolation: Isolating the experimental setup from environmental noise (*e.g., temperature control, vibration damping*) can minimize decoherence.
- Error Correction Protocols: Implementing quantum error correction protocols can help preserve entanglement in the presence of decoherence and measurement errors.

Decoherence and measurement issues present significant challenges in photon-based entanglement experiments. Decoherence transforms pure states into mixed states, degrading entanglement, while measurement limitations, such as detector inefficiency and dark counts, introduce uncertainties and potential biases. Addressing these issues through advanced technology and careful experimental design is crucial for accurately studying quantum entanglement.

2. Theoretical Considerations:

Foundations of Quantum Mechanics

The foundations of quantum mechanics are a subject of ongoing debate, especially regarding the interpretation of quantum phenomena and the nature of reality they describe. Different interpretations offer varying perspectives on what quantum entanglement means and how it should be understood. We will explore several key interpretations and their implications for entanglement.

Copenhagen Interpretation

Overview:

- Proposed by Niels Bohr and Werner Heisenberg, the Copenhagen interpretation is one of the oldest and most widely taught interpretations of quantum mechanics.
- It posits that a quantum system exists in a superposition of states until it is observed or measured, at which point it collapses to a definite state.

Implications for Entanglement:
- According to this interpretation, entangled particles remain in a superposition of states until a measurement collapses their wavefunction, instantaneously determining the state of both particles.
- The phenomenon of wavefunction collapse implies that the act of measurement affects the system, introducing an element of randomness.

Mathematical Expression:
- The state of an entangled system of two particles can be described by a wavefunction $|\psi\rangle$,

$$|\psi\rangle = \alpha|0\rangle_A|1\rangle_B + \beta|1\rangle_A|0\rangle_B$$

[10.0]

where α and β are complex coefficients, and $|0\rangle$ and $|1\rangle$ represent the basis states of particles A and B.

Many-Worlds Interpretation

Overview:
- Proposed by Hugh Everett III, the many-worlds interpretation suggests that all possible outcomes of a quantum measurement actually occur, each in a separate, branching universe.
- There is no wavefunction collapse; instead, the universe splits into multiple, non-interacting branches.

Implications for Entanglement:
- In this interpretation, entangled particles do not collapse into a single state upon measurement. Instead, each possible measurement outcome exists in a different branch of the universe.
- Entanglement is understood as a correlation between particles across these parallel universes.

Mathematical Expression:

- The universal wavefunction evolves according to the Schrödinger equation without collapse

$$|\Psi(t)\rangle = \sum_i c_i(t)|\phi_i\rangle$$

[11.0]

where $c_i(t)$ are time-dependent coefficients and $|\phi_i\rangle$ are the possible states of the system. Each term in the sum represents a different branch of the universe.

Objective Collapse Theories

Overview:

- Objective collapse theories, such as the Ghirardi-Rimini-Weber (GRW) theory, propose that wavefunction collapse is a spontaneous, independent-of-observation physical process.
- These theories aim to modify the standard quantum mechanics framework to include a mechanism for collapse.

Implications for Entanglement:

- According to objective collapse theories, entangled states can spontaneously collapse, leading to definite outcomes for measurements.
- This introduces a new element of randomness and locality to the process of entanglement and measurement.

Mathematical Expression:

- The GRW theory modifies the Schrödinger equation by adding a nonlinear, stochastic term to describe spontaneous collapses

$$d|\psi\rangle = \left(-\frac{i}{\hbar}H\,dt + \sum_k \left(L_k - \frac{1}{2}L_k^\dagger L_k dt\right)\right)|\psi\rangle$$

[12.0]

where L_k are collapse operators.

Relational Quantum Mechanics

Overview:

- Proposed by Carlo Rovelli, relational quantum mechanics suggests that the state of a quantum system is relative to the observer and the system being measured.
- There is no absolute state of the system; states are only meaningful in relation to other systems.

Implications for Entanglement:

- Entanglement, in this interpretation, is understood as a correlation between systems relative to an observer.
- Measurement outcomes depend on the observer-system relationship, and different observers may have different accounts of the same quantum event.

Mathematical Expression:

- The state of a system relative to an observer O can be represented as

$$\rho_{S|O} = \sum_i p_i |\psi_i\rangle_S \langle\psi_i|$$

[13.0]

where $\rho_{S|O}$ is the density matrix of the system S relative to the observer O.

De Broglie-Bohm Theory

Overview:

- Also known as the pilot-wave theory, proposed by Louis de Broglie and David Bohm, this interpretation posits that particles have definite positions and velocities guided by a *"pilot wave."*
- The wavefunction provides the information that determines the trajectories of particles.

Implications for Entanglement:

- Entangled particles have well-defined properties, and their correlations are determined by the guiding wave.
- This interpretation introduces hidden variables that explain the outcomes of quantum measurements.

Mathematical Expression:

- The evolution of the wavefunction and particle positions is given by:

$$\frac{d\mathbf{r}_i}{dt} = \frac{\hbar}{m_i} \text{Im} \left(\frac{\nabla_i \psi(\mathbf{r}_1, \mathbf{r}_2, ..., \mathbf{r}_N, t)}{\psi(\mathbf{r}_1, \mathbf{r}_2, ..., \mathbf{r}_N, t)} \right)$$

[14.0]

where $\mathbf{r}_i$ are the positions of the particles.

The interpretation of quantum mechanics significantly influences our understanding of entanglement. The Copenhagen interpretation emphasizes the role of measurement, the many-worlds interpretation eliminates collapse in favor of branching universes, objective collapse theories introduce physical collapse mechanisms, relational quantum mechanics highlights the observer-system relationship, and de Broglie-Bohm theory provides a deterministic framework with hidden variables. Each interpretation offers unique insights and challenges, contributing to the rich and ever-evolving debate on the foundations of quantum mechanics. This debate is as relevant and timely as ever.

String Theory's Potential

String Theory and Its Framework

String theory proposes that the fundamental constituents of the universe are not point-like particles but rather one-dimensional "*strings*" that vibrate at specific frequencies. These vibrations determine the properties of particles, such as mass, charge, and spin. String theory offers a broader and potentially deeper framework for understanding fundamental interactions in physics, providing insights that extend beyond the Standard Model of particle physics.

Key Features of String Theory:

- Vibrational Modes: Strings' vibrational modes correspond to different particles. The same string can represent different particles depending on how it vibrates.
- Extra Dimensions: String theory requires additional spatial dimensions beyond the familiar three. Typically, it suggests 10 or 11 dimensions, with the extra dimensions compactified at small scales.
- Supersymmetry: Many string theory formulations include supersymmetry, a symmetry that relates bosons (*force-carrying particles*) and fermions (*matter particles*).

Mathematical Framework

String Action:

- The action of a string, analogous to the action in point-particle mechanics, is given by the Nambu-Goto action

$$S = -T \int d\tau d\sigma \sqrt{-\gamma}$$

[15.0]

where T is the string tension, τ and σ are the worldsheet coordinates (*parameters describing the surface swept by the string*), and γ is the determinant of the induced metric on the string worldsheet.

Equations of Motion:

- The dynamics of strings are described by the Polyakov action, leading to the equations of motion for the string coordinates $X^\mu(\tau, \sigma)$

$$\frac{\partial}{\partial \tau}\left(\sqrt{-\gamma}\gamma^{ab}\frac{\partial X^\mu}{\partial \sigma^a}\right) = 0$$

[16.0]

where γ^{ab} is the inverse metric on the worldsheet, and σ^a are the worldsheet coordinates.

Implications for Entanglement

String theory suggests additional degrees of freedom and characteristics that could be relevant for understanding quantum entanglement:

2.1. Vibrational Modes and Entanglement:
- Particles represented by strings vibrating in identical or harmonically related modes may exhibit strong correlations. A string's vibrational state can influence a particle's quantum state, potentially leading to entanglement when strings share similar vibrational characteristics.

Wavefunction Representation: The quantum state of a string can be described by a wavefunction $\psi(X)$ that incorporates the vibrational modes. For two entangled strings, the combined wavefunction might be

$$\Psi(X_1, X_2) = \sum_n \psi_n(X_1)\phi_n(X_2)$$

[17.0]

where X_1 and X_2 represent the coordinates of the two strings, and ψ_n and ϕ_n are vibrational mode functions.

2.2. Extra Dimensions and Entanglement:
- The presence of extra dimensions can influence how strings interact and entangle. Strings vibrating in higher dimensions might experience entanglement mechanisms not present in lower-dimensional models.
- Compactification: The compactification of extra dimensions can lead to different effective potentials and interactions, potentially enhancing or modifying entanglement properties.

2.3. Gauge Symmetries and Entanglement:
- String theory incorporates various gauge symmetries that govern the interactions of strings. Shared gauge symmetries between strings can lead to correlated quantum states, resulting in entanglement.

Yang-Mills Fields: The interactions of strings can be described using gauge fields, leading to equations similar to those in Yang-Mills theory

$$\mathcal{L} = -\frac{1}{4} F^a_{\mu\nu} F^{\mu\nu a}$$

[18.0]

where $F^a_{\mu\nu}$ the field strength tensors for the gauge fields are.

2.4. String Tension and Coupling Constants:
- The tension of a string, which is a fundamental parameter in string theory, affects its vibrational modes and interactions. Strings with similar tension values may exhibit enhanced entanglement due to shared energy characteristics.
- Coupling Constants: The coupling constants in string interactions determine the strength of entanglement. Strings with identical or harmonically related coupling constants may exhibit stronger entanglement.

By considering additional degrees of freedom, such as vibrational modes, extra dimensions, gauge symmetries, and string tension, we can uncover new mechanisms and principles governing entanglement. This broader perspective may address limitations in current photon-based experiments and provide deeper insights into the fundamental nature of quantum mechanics.

3. Innovative Perspectives:

Expanding the Scope of Entanglement

By considering additional characteristics like vibrational modes, string tension, and gauge symmetries, we can explore a more general form of entanglement that might apply to a wider range of particles and conditions. This approach leverages the theoretical framework provided by string theory and other advanced quantum theories to broaden our understanding of entanglement beyond traditional photon-based experiments.

Vibrational Modes

Concept:
- In string theory, different vibrational modes of a string correspond to different particle types and properties. These vibrational modes can be thought of as the fundamental frequencies at which strings oscillate.
- Shared vibrational modes between particles can lead to entanglement, as the particles share a common underlying quantum state.

Mathematical Representation:
- The quantum state of a string can be represented as a superposition of its vibrational modes:

$$|\psi\rangle = \sum_n c_n |n\rangle$$

[19.0]

where $|n\rangle$ represents the n-th vibrational mode and c_n are the corresponding coefficients.

- For two entangled strings, the combined state might be:

$$|\Psi\rangle = \sum_{m,n} c_{mn} |m\rangle_1 |n\rangle_2$$

[19.1]

where $|m\rangle_1$ and $|n\rangle_2$ are the vibrational modes of the first and second strings, respectively, and c_{mn} are the coefficients that describe the entanglement.

String Tension

Concept:
- The tension of a string, denoted by T, is a fundamental parameter that influences its vibrational energy. Strings with identical or harmonically related tensions may exhibit stronger entanglement due to their shared energy properties.

Mathematical Representation:
- The energy of a vibrating string is proportional to its tension T and its vibrational frequency ν

$$E \propto T\nu^2$$

[20.0]

- Two strings with tensions T_1 and T_2 might be entangled if their tensions are related by a simple ratio, enhancing their correlated quantum states:

$$\frac{T_1}{T_2} =$$

[20.1]

where k is a rational number.

Gauge Symmetries

Concept:
- Gauge symmetries are symmetries of the fields that describe fundamental forces. In string theory, different particles arise as excitations of strings under various gauge symmetries.
- Shared gauge symmetries between strings can lead to entanglement, as the strings are governed by the same fundamental interactions.

Mathematical Representation:
- The Lagrangian for a gauge field is given by:

$$\mathcal{L} = -\frac{1}{4} F^a_{\mu\nu} F^{\mu\nu a}$$

[21.0]

where $F^a_{\mu\nu}$ are the field strength tensors for the gauge fields.

- For two strings with gauge symmetry groups G_1 and G_2, entanglement may arise if

$$G_1 \cong G_2$$

[21.1]

indicates that the gauge groups are isomorphic, leading to correlated quantum states.

Dimensional Embedding

Concept:

- String theory posits that strings exist in higher-dimensional spaces. The specific dimensions and orientations of these strings can influence their interactions and potential for entanglement.

Mathematical Representation:

- The coordinates of a string in a higher-dimensional space can be described by:

$$X^\mu(\tau, \sigma), \quad \mu = 0, 1, 2, \ldots, D-1$$

[22.0]

where D is the number of dimensions, and τ and σ are the worldsheet coordinates.

- Two strings may be entangled if their embeddings in the higher-dimensional space share certain geometric or topological properties:

$$\text{Entangled if: } f_1(X^\mu) = f_2(X^\mu)$$

[22.1]

where f_1 and f_2 are functions describing the embeddings of the two strings.

Practical Implications and Experiments (expand on this later)

3.1. Quantum Dot Arrays:
- Quantum dots can simulate vibrational modes and other properties of strings. By creating arrays of quantum dots with controlled vibrational modes and interactions, researchers can study the resulting entanglement.

3.2. Superconducting Qubits:
- Superconducting qubits with engineered properties can mimic the behaviors of strings under various tensions and gauge symmetries. Experiments with these qubits can provide insights into the entanglement mechanisms predicted by string theory.

3.3. High-Energy Physics Experiments:
- Particle accelerators can generate particles with specific string characteristics. By analyzing the resulting particles and their interactions, researchers can test the predictions of string theory regarding entanglement.

Expanding the scope of entanglement to include characteristics such as vibrational modes, string tension, gauge symmetries, and dimensional embedding allows us to explore a more general and comprehensive form of entanglement. This approach leverages the rich theoretical framework of string theory and other advanced quantum theories, potentially revealing new mechanisms and principles governing quantum entanglement across a wider range of particles and conditions. This expanded perspective integrates theoretical insights and practical approaches by highlighting the potential for discoveries in quantum entanglement through the lens of string theory.
Overcoming Limitations

Addressing Potential Flaws in Photon Experiments

Photon-based experiments have been crucial in demonstrating quantum entanglement and testing the foundations of quantum mechanics. However, they also face specific limitations that can obscure our understanding of entanglement. By exploring alternative characteristics and methods, we can address these potential flaws and gain new insights into quantum mechanics.

Limitations in Photon Experiments

3.4.1. Decoherence:
- Interaction with Environment: Photons are susceptible to interactions with their environment, which can lead to decoherence. This interaction causes the quantum superposition states to collapse into classical states, thereby destroying entanglement.

Mathematical Representation: The evolution of a quantum state ρ under decoherence can be modeled by a master equation

$$\frac{d\rho}{dt} = -\frac{i}{\hbar}[H, \rho] + \sum_k \left(L_k \rho L_k^\dagger - \frac{1}{2}\{L_k^\dagger L_k, \rho\} \right)$$

[23.0]

where L_k are Lindblad operators representing the decoherence processes.

3.4.2. Measurement Issues:

- Detector Inefficiency: Photon detectors often have less than perfect efficiency, leading to missed detections that can skew experimental results.
- Dark Counts: False positives in photon detection, known as dark counts, can introduce errors in the measurement of entangled states.
- Mathematical Representation: The probability P_{det} of detecting a photon can be expressed as

$$P_{det} = \eta P_{photon}$$

[24.0]

where η is the detection efficiency.

3.4.3. Loopholes in Bell Tests:

- Locality Loophole: Ensuring that measurements are space-like separated is challenging, leading to potential causal connections that could explain the observed correlations classically.
- Detection Loophole: Incomplete detection of entangled particles can result in biased samples, leading to inaccurate conclusions about entanglement.
- Bell's Inequality: The violation of Bell's inequality, which tests the presence of entanglement, relies on measuring correlations $(E(\hat{a}, \hat{b})$ between different settings:

$$S = |E(\hat{a}, \hat{b}) - E(\hat{a}, \hat{b}') + E(\hat{a}', \hat{b}) + E(\hat{a}', \hat{b}')| \leq 2$$

[25.0]

Alternative Characteristics for Testing Entanglement

3.5.1. Vibrational Modes in Strings:
- String Theory: Considering vibrational modes of strings can provide a new perspective on entanglement. Strings vibrating at specific frequencies may exhibit entanglement if their modes are harmonically related.
- Mathematical Expression: The wavefunction of an entangled state of two strings might be represented as

$$|\Psi\rangle = \sum_{m,n} c_{mn} |m\rangle_1 |n\rangle_2$$

[26.0]

where $|m\rangle$ and $|n\rangle$ are the vibrational modes.

3.5.2. Spin States and Spin Chains:
- Spin Chains: Systems of interacting spins (*e.g., in magnetic materials or quantum dots*) can exhibit entanglement that can be studied through their spin states.
- Heisenberg Model: The Hamiltonian for a simple spin chain can be written as:

$$H = J \sum_{i} \mathbf{S}_i \cdot \mathbf{S}_{i+1}$$

[26.1]

where J is the exchange interaction and $\mathbf{S}_i$ are the spin operators.

3.5.3. Topological States and Quantum Hall Effect:
- Topological Insulators: Certain materials exhibit topologically protected states that are robust against local perturbations, which can be used to study entanglement.
- Chern Number: The topological properties can be characterized by a Chern number C, which is an integer representing the topological invariant of the system.

3.5.4. Superconducting Qubits:
- **Cooper Pairs**: Superconducting qubits, which involve Cooper pairs of electrons, can be used to study entanglement through the Josephson junctions and flux qubits.
- **Hamiltonian**: The Hamiltonian for a superconducting qubit can be expressed as:

$$H = 4E_C(n - n_g)^2 - E_J \cos(\phi)$$

[27.0]

where E_C is the charging energy, E_J is the Josephson energy, n is the number of Cooper pairs, n_g is the gate charge, and ϕ is the phase difference across the junction.

Overcoming Measurement Issues

3.6.1. High-Efficiency Detectors:
- **SNSPDs**: Superconducting nanowire single-photon detectors offer high efficiency and low dark counts, which can significantly improve the accuracy of photon measurements.
- **Mathematical Impact**: Improved detection efficiency reduces the probability of missing entangled photons:

$$P_{det} = \eta_{high} P_{photon}$$

[28.0]

where η_{high} represents the improved efficiency.

3.6.2. Quantum Error Correction:
- **Error Correction Codes**: Implementing quantum error correction codes can help preserve entanglement in the presence of decoherence and measurement errors.
- **Stabilizer Codes**: An Case is the use of stabilizer codes, where the logical qubits are protected by encoding into multiple physical qubits.

By addressing the limitations in photon-based entanglement experiments through alternative characteristics and improved measurement techniques, we can gain deeper insights into the nature of quantum entanglement. Exploring vibrational modes, spin states, topological properties, and superconducting qubits provides a broader framework for studying entanglement, potentially resolving ambiguities in current quantum mechanics.

Detailed Hypothesis

Hypothesis: Particles achieve quantum entanglement if they share two or more fundamental characteristics derived from string theory, such as vibrational modes, temporal synchronization, string tension, topological properties, dimensional embedding, gauge symmetries, and coupling constants. These shared characteristics establish a quantum state that correlates the particles, enabling entanglement even across vast distances.

Key Characteristics for Entanglement

1. Vibrational Modes:

Concept:
- Strings vibrate at specific frequencies, which determine the properties of the particles they represent. When two or more strings share the same or harmonically related vibrational modes, they can achieve a common quantum state, potentially leading to entanglement.

Mathematical Representation:
- The state of a string can be described by its vibrational modes. For an open string, these modes are quantized and can be expressed in terms of creation and annihilation operators acting on a vacuum state $|0\rangle$:

$$|n\rangle = a_n^\dagger |0\rangle$$

[29.0]

where $a_n^\dagger$ is the creation operator for the n-th vibrational mode.

- For two strings with vibrational modes m and n, the entangled state can be written as

$$|\Psi\rangle = \sum_{m,n} c_{mn} |m\rangle_1 |n\rangle_2 \quad |\Psi\rangle = \sum_{m,n} c_{mn} |m\rangle_1 |n\rangle_2$$

[29.1]

where $|m\rangle_1$ and $|n\rangle_2$ are the vibrational modes of the first and second strings, respectively, and c_{mn} are coefficients representing the degree of entanglement between these modes.

Implications for Entanglement:
- **Shared Modes:** If two strings share identical or harmonically related vibrational modes, their quantum states can become correlated. This correlation can be described by a shared wave function that encapsulates the entangled state.

- Energy and Frequency Matching: The energy of the vibrational modes, given by $E_n = \hbar\omega_n$, where ω_n is the frequency of the n-th mode, must match or harmonize for entanglement to occur.

Case:
- Consider two strings, each with fundamental frequencies ω_1 and ω_2. If $\omega_1 = k\omega_2$ (where k is an integer), the strings can share harmonically related vibrational modes, potentially leading to entanglement:

$$\omega_n = n\omega_1 = nk\omega_2 \quad \omega_n = n\omega_1 = nk\omega_2$$

[29.2]

Temporal Synchronization

Concept:
- Temporal synchronization refers to the alignment of the phase or timing of quantum states. When particles are synchronized in time, their quantum states can become entangled due to the temporal coherence.

Mathematical Representation:
- Temporal synchronization can be described by a phase factor $\phi(t)$ that affects the quantum state:

$$|\psi(t)\rangle = e^{i\phi(t)}|\psi(0)\rangle$$

[30.0]

For two synchronized particles, their combined state can be written as

$$|\Psi(t)\rangle = e^{i(\phi_1(t)+\phi_2(t))}|\Psi(0)\rangle$$

[30.1]

Implications for Entanglement:
- Phase Coherence: Maintaining phase coherence over time is crucial for entanglement. External disturbances that disrupt this coherence can destroy the entangled state.
- Synchronization Mechanisms: Utilizing technologies such as synchronized lasers or atomic clocks can help achieve the necessary temporal alignment for entanglement.

String Tension

Concept:
- The tension of a string, denoted by T, affects its vibrational properties and energy levels. Strings with the same or harmonically related tensions can achieve correlated quantum states.

Mathematical Representation:
- The energy of a string's vibrational mode is proportional to its tension T and its frequency ω:

$$E \propto T\omega^2$$

[31.0]

Implications for Entanglement:
- Tension Matching: For two strings to become entangled, their tensions must be related in a way that allows their vibrational modes to harmonize. This can be expressed as

$$T_1 = kT_2$$

[31.1]

where k is a rational number.

Topological Properties

Concept:
- Strings can have different topological configurations, such as being open or closed. The topological properties of strings influence their interactions and potential for entanglement.

Mathematical Representation:
- The topology of a string can be described by its boundary conditions. For an open string, the ends are free, while for a closed string, the ends are joined:

$$X^\mu(\tau, \sigma) = X^\mu(\tau, \sigma + 2\pi)$$

[32.0]

Implications for Entanglement:
- Topological Compatibility: Strings with compatible topological properties are more likely to entangle. For Case, two closed strings with similar winding numbers can achieve a common quantum state.

Dimensional Embedding

Concept:
- Strings exist in higher-dimensional spaces, and their embedding in these dimensions affects their vibrational properties and interactions.

Mathematical Representation:
- The coordinates of a string in higher dimensions are given by

$$X^\mu(\tau, \sigma), \quad \mu = 0, 1, 2, \ldots, D-1$$

[33.0]

where D is the number of dimensions.

Implications for Entanglement:
- Dimensional Alignment: Strings embedded in the same or harmonically related dimensions can achieve entanglement. The functions describing their embeddings $f_1(X^\mu)$ and $f_2(X^\mu)$ must be related.

Gauge Symmetries

Concept:
- Gauge symmetries govern the interactions of strings. Shared gauge symmetries between strings can lead to entanglement.

Mathematical Representation:
- The Lagrangian for a gauge field is

$$\mathcal{L} = -\frac{1}{4} F^a_{\mu\nu} F^{\mu\nu a}$$

[34.0]

where $F^a_{\mu\nu}$ are the field strength tensors.

Implications for Entanglement:
- Symmetry Matching: Strings with isomorphic gauge symmetry groups can achieve entanglement:

$$G_1 \cong G_2$$

[34.1]

Coupling Constants

Concept:
- The strength of interactions between strings is determined by coupling constants. Strings with similar coupling constants may exhibit correlated quantum states.

Mathematical Representation:
- The interaction strength can be described by a coupling constant g:

$$H_{int} = g \int d\sigma \, J(\sigma)$$

[35.0]

Implications for Entanglement:
- Coupling Alignment: Strings with coupling constants related by a simple ratio can achieve stronger entanglement:

$$g_1 = k g_2$$

[35.1]

2. Temporal Synchronization:

Concept

Temporal synchronization refers to the alignment of the phases or timings of quantum states of particles. When particles are temporally synchronized, their quantum states can become entangled due to the coherence established in their temporal alignment. This synchronization can occur naturally or be induced through external manipulation, such as synchronized laser pulses or atomic clocks.

Importance in Quantum Entanglement

Temporal synchronization is crucial in quantum entanglement because it ensures that particles' quantum states are coherent over time. This coherence is essential for maintaining the entanglement, as any discrepancy in timing can lead to decoherence and the loss of the entangled state.

Mathematical Representation

2.1. Quantum States and Phase Coherence:
- - The quantum state of a particle can be represented as

$$|\psi(t)\rangle = e^{i\phi(t)}|\psi(0)\rangle$$

[36.0]

where ϕ(t) is the phase factor that evolves. For two particles to be temporally synchronized, their phase factors must be aligned.

2.2. Combined State of Synchronized Particles:
- For two synchronized particles, the combined quantum state can be written as

$$|\Psi(t)\rangle = e^{i(\phi_1(t)+\phi_2(t))}|\Psi(0)\rangle$$

[36.1]

Here, $\phi_1(t)$ and $\phi_2(t)$ are the phase factors for the first and second particles, respectively.

2.3. Entanglement and Temporal Synchronization:
- When particles are entangled, their quantum states are interdependent. If particles are temporally synchronized, the entangled state can be described as

$$|\Psi(t)\rangle = \alpha e^{i\phi(t)}|0\rangle_1|1\rangle_2 + \beta e^{-i\phi(t)}|1\rangle_1|0\rangle_2$$

[37.0]

where α and β are the coefficients representing the probability amplitudes and ϕ(t) the synchronized phase factor.

2.4. Bell's Inequality and Temporal Synchronization:
- Temporal synchronization is essential in Bell test experiments, where measurements are made at specific time intervals. The correlation function $E(\hat{a}, \hat{b}, t)$ can depend on the synchronization of measurements:

$$E(\hat{a}, \hat{b}, t) = \langle\Psi(t)|\hat{a} \otimes \hat{b}|\Psi(t)\rangle$$

[38.0]

Ensuring that the measurements are temporally synchronized is crucial for accurately testing Bell's inequalities.

Practical Implementation

2.5.1. Synchronized Lasers:
- Lasers can be used to synchronize particles' quantum states. By sending synchronized laser pulses to particles, their phases can be aligned, ensuring temporal coherence.
- Case: Ultrafast lasers can emit pulses with precise timing, used to excite particles simultaneously, thereby achieving temporal synchronization.

2.5.2. Atomic Clocks:
- Atomic clocks, known for their precision, can synchronize particles' quantum states by providing a common reference time. This ensures that the particles' phase factors evolve coherently.
- Case: Cesium or rubidium atomic clocks can be used in quantum experiments to provide synchronized timing signals.

2.5.3. Entangled Photon Sources:
- Spontaneous parametric down-conversion (SPDC) can generate entangled photons that are temporally synchronized. By ensuring the timing of photon pair generation is precise, temporal synchronization can be maintained.
- Case: In SPDC, a pump laser is used to create entangled photon pairs with synchronized emission times, ensuring their phases are aligned.

Experimental Considerations

2.6.1. Decoherence Mitigation:
- To maintain temporal synchronization, it is crucial to minimize decoherence caused by environmental factors such as thermal fluctuations or electromagnetic interference. Isolating the experimental setup and using noise-canceling techniques can help.
- Case: Conducting experiments in cryogenic environments can reduce thermal noise and preserve temporal coherence.

2.6.2. Precision Measurement:
- High-precision measurement devices are necessary to detect and maintain temporal synchronization. Detectors with low-timing jitter and high resolution are essential for accurate phase alignment.
- Case: Single-photon avalanche diodes (SPADs) with low timing jitter can be used to measure the arrival times of photons with high precision.

3. String Tension:

Concept

String tension is a fundamental parameter in string theory that influences the vibrational properties and energy levels of strings. The tension, denoted by T, determines how strings vibrate and interact with other strings. When two or more strings have identical or harmonically related tensions, their vibrational energies can become correlated, facilitating entanglement due to their shared energetic properties.

Importance in Quantum Entanglement

Identical or harmonically related string tensions can lead to entanglement because the vibrational modes of the strings become synchronized. This synchronization means that the energy states of the strings are interdependent, allowing the quantum states of the strings to become entangled.

Mathematical Representation

3.1. String Tension and Vibrational Energy:
- The tension T in a string affects its vibrational frequency ω and energy E. For a string of length L, the fundamental frequency ω is given by

$$\omega = \frac{n\pi}{L}\sqrt{\frac{T}{\mu}}$$

[39.0]

where n is a mode number and μ is the linear mass density of the string.
- The energy associated with the vibrational mode \(n\) can be expressed as

$$E_n = \hbar\omega_n = \hbar\frac{n\pi}{L}\sqrt{\frac{T}{\mu}}$$

[39.1]

3.2. Correlated Energies and Entanglement:

- For two strings to become entangled, their tensions T_1 and T_2 must be related such that their vibrational frequencies are harmonically related. This can be represented as

$$T_1 = kT_2$$

[40.0]

where k is a rational number.

- The combined state of two entangled strings with tensions T_1 and T_2 can be written as a superposition of their vibrational modes:

$$|\Psi\rangle = \sum_{m,n} c_{mn} |m\rangle_{T_1} |n\rangle_{T_2}$$

[40.1]

where $|m\rangle_{T_1}$ and $|n\rangle_{T_2}$ represent the vibrational states of the first and second strings, respectively, and c_{mn} are coefficients representing the entanglement.

Implications for Entanglement

3.2.1. Energy Matching and Synchronization:

- When strings have identical or harmonically related tensions, their vibrational energies become matched, leading to the synchronization of their quantum states. This synchronization is essential for maintaining coherence and entanglement.

- Case: Consider two strings with tensions $T_1 = 2T_2$. Their fundamental frequencies ω_1 and ω_2 will be related as $\omega_1 = 2\omega_2$, ensuring that their vibrational modes can harmonize.

3.2.2. String Coupling and Interaction:
- The interaction between strings with related tensions can enhance entanglement through coupling mechanisms. The coupling constant g for the interaction can be influenced by the string tensions:

$$H_{int} = g \int d\sigma \, J(\sigma)$$

[41.0]

where $J(\sigma)$ is the current density along the string.

Practical Implementation

3.3.1. Experimental Setup for String Tension Matching:
- To achieve entanglement through string tension, experimental setups must ensure precise control over the string tensions. This can be done using mechanical or electromagnetic tuning methods.

- Case: In a lab, strings (or analogous systems like superconducting circuits) can be stretched or compressed to achieve the desired tension, harmonizing their vibrational modes.

3.3.2. Measurement Techniques:

Concept:

Measuring the vibrational modes and energies of strings requires high-precision instruments capable of detecting minute displacements and quantum states. Techniques such as laser interferometry and quantum state tomography are critical for analyzing these vibrational states and confirming entanglement.

Importance in Quantum Entanglement

High-precision measurement techniques are essential for validating the presence and characteristics of entanglement in particles with shared string properties. These techniques provide the necessary data to compare experimental results with theoretical predictions, ensuring accurate confirmation of entanglement.

Detailed Measurement Techniques

1. Laser Interferometry:
- Concept: Laser interferometers measure small displacements by detecting phase shifts in laser beams. When a laser beam is split and travels along two different paths, any difference in the lengths of these paths causes an interference pattern when the beams are recombined. This interference pattern can be analyzed to measure minute displacements and vibrations.

Mathematical Framework:
- The phase difference Δϕ between the two paths is related to the path length difference ΔL.

$$\Delta\phi = \frac{2\pi}{\lambda}\Delta L$$

[41.1]

where λ is the wavelength of the laser light.

- The interference intensity I as a function of the phase difference is given by

$$I(\Delta\phi) = I_0\left(1 + \cos(\Delta\phi)\right)$$

[41.2]

where I_0 is the maximum intensity.

- Case: Laser interferometers can measure the minute displacements and vibrations of strings, providing data on their vibrational frequencies and energies. This data can be used to analyze the string's vibrational modes and confirm entanglement.

2. Quantum State Tomography:
- Concept: Quantum state tomography reconstructs the quantum state of a system by measuring various properties and using statistical methods to infer the state. This process involves performing a series of measurements on a quantum system and reconstructing the density matrix that represents the quantum state.

Mathematical Framework:
- The density matrix ρ of a quantum state can be reconstructed from measurement data

$$\rho = \sum_{i,j} \rho_{ij} |i\rangle \langle j|$$

[41.2]

- To reconstruct ρ, measurements in different bases are required. The probability of measuring outcome m in basis b is given by

$$P(m|b) = \text{Tr}(\rho M_{m|b})$$

[41.3]

where $M_{m|b}$ is the measurement operator.

- Case: Quantum state tomography can reconstruct the entangled state of a system, providing detailed information about the correlation induced by similar vibrational modes and other string characteristics.

3. Spectroscopic Analysis:

- Concept: Spectroscopy involves studying the interaction between matter and electromagnetic radiation to infer the properties of particles. It is used to measure energy levels, transition frequencies, and other properties that provide insights into the coupling constants and the structure of quantum states.

Mathematical Framework:

- The energy levels E_n of a system can be related to the observed spectral lines

$$E_n = h\nu_n$$

[41.3.1]

where h is Planck's constant and ν_n is the frequency of the spectral line.

- The intensity of spectral lines can provide information about the population of energy states and transition probabilities.
- Case: In particle accelerators, spectroscopic analysis can be used to measure the energy levels and transition frequencies of generated particles, helping to determine the coupling constants.

4. Interferometric Measurement of String Tension:
    - Concept: The tension in a string affects its vibrational energy, which can be measured using interferometric techniques. These measurements can provide data on the string's vibrational modes and the energy associated with different tensions.

Mathematical Framework:
- The energy E_n of a vibrating string mode n with tension T) is given by,

$$E_n = n\hbar\omega_n = n\hbar\sqrt{\frac{T}{\mu}}$$

[41.3.2]

where ω_n is the angular frequency of the n-th mode and μ is the linear mass density of the string.

- Case: By measuring the vibrational frequencies ω_n using laser interferometry, the tension T in the string can be determined. This data can be correlated with other measurements to confirm entanglement.

Innovative Approaches

1. Integrated Quantum Sensors:
   - Develop integrated quantum sensors that combine elements of laser interferometry, quantum state tomography, and spectroscopic analysis. These sensors can provide comprehensive measurements of quantum states with high precision.
   - Case: A hybrid quantum sensor using NV centers in diamond can perform quantum state tomography and spectroscopic analysis simultaneously, providing detailed information about the quantum state and its environment.

2. Machine Learning for Quantum Measurements:
   - Use machine learning algorithms to analyze measurement data and improve the accuracy of quantum state reconstruction. These algorithms can handle large datasets and identify patterns that traditional methods might miss.

Mathematical Framework:
   - Machine learning models can be trained to predict the density matrix ρ from measurement outcomes:

$$\hat{\rho} = f(\{P(m|b)\}; \theta)$$

[41.4]

where f is a machine learning model parameterized by θ.

   - Case: Neural networks can be trained to reconstruct quantum states from tomography data, providing faster and more accurate state determination.

3. Adaptive Measurement Techniques:
- Implement adaptive measurement techniques that dynamically adjust the measurement parameters based on real-time data. This approach optimizes the measurement process and improves the quality of the reconstructed quantum states.
- Case: Adaptive quantum state tomography uses feedback from initial measurements to refine the subsequent measurement settings, reducing the number of measurements needed for accurate state reconstruction.

High-precision measurement techniques are essential for accurately determining the vibrational modes and energies of strings, confirming the presence of entanglement. Laser interferometry, quantum state tomography, and spectroscopic analysis provide robust methods for measuring these properties. Innovative approaches, including integrated quantum sensors, machine learning for quantum measurements, and adaptive measurement techniques, offer enhanced capabilities for detecting and understanding quantum entanglement in complex systems. Importantly, these advancements hold the potential to significantly deepen our understanding of quantum mechanics and accelerate the development of practical quantum technologies.

String tension plays a critical role in determining the vibrational properties and energy levels of strings. When strings have identical or harmonically related tensions, their vibrational energies can become synchronized, leading to entanglement. This synchronization allows the quantum states of the strings to be correlated, enabling entanglement across vast distances. Experimental setups that control string tension precisely, combined with advanced measurement techniques, can provide new insights into the nature of quantum entanglement through string theory.

4. Topological Properties:

Concept

In string theory, the topological properties of strings, such as whether they are open or closed, significantly influence their interactions and potential for entanglement. Topological configurations refer to the way strings are shaped and connected, which can affect their vibrational modes and how they interact with other strings. Strings with similar topological properties can exhibit correlated behaviors that lead to entanglement.

Importance in Quantum Entanglement

The topology of a string, whether open or closed, determines the boundary conditions and the types of vibrational modes the string can support. Strings with similar topological configurations can interact in ways that synchronize their quantum states, facilitating entanglement. This is because the vibrational modes and allowed interactions are dictated by the string's topology.

Mathematical Representation

4.1. Open Strings:
- An open string has two endpoints, and its boundary conditions are typically

$$X^\mu(\tau, \sigma = 0) = X^\mu(\tau, \sigma = \pi)$$

[42.0]

where X^μ represents the position of the string in spacetime, τ is the worldsheet time coordinate, and σ is the spatial coordinate along the string.

- The vibrational modes of an open string can be expressed as

$$X^\mu(\tau, \sigma) = x^\mu + 2\alpha' p^\mu \tau + i\sqrt{2\alpha'} \sum_{n \neq 0} \frac{1}{n} \alpha_n^\mu \cos(n\sigma) e^{-in\tau}$$

[42.1]

where α' is the string tension parameter, p^μ is the momentum, and α_n^μ are the mode operators.

4.2. Closed Strings:
- A closed string forms a loop with no endpoints, and its boundary conditions are:

$$X^\mu(\tau, \sigma) = X^\mu(\tau, \sigma + 2\pi)$$

[43.0]

- The vibrational modes of a closed string can be written as:

$$X^\mu(\tau, \sigma) = x^\mu + 2\alpha' p^\mu \tau + i\sqrt{\alpha'} \sum_{n \neq 0} \frac{1}{n} \left(\alpha_n^\mu e^{-in(\tau-\sigma)} + \tilde{\alpha}_n^\mu e^{-in(\tau+\sigma)} \right)$$

[44.0]

where $alpha_n^\mu$ and $\tilde{\alpha}_n^\mu$ are the left-moving and right-moving mode operators, respectively.

Implications for Entanglement

4.2.1 Boundary Conditions and Mode Synchronization:
- The boundary conditions of open and closed strings lead to different sets of vibrational modes. When two strings share the same topology, their modes can synchronize, leading to entanglement.
- Case: Two open strings with identical boundary conditions can exhibit synchronized vibrational modes if they are subjected to the same external influences.

4.2.2. Interaction Channels:
- Similar topological properties allow strings to interact through specific channels. For Case, closed strings can interact through loops, leading to entangled states if their vibrational modes are harmonized.
- Interaction Hamiltonian: The Hamiltonian governing the interaction of two strings with similar topology can be expressed as

$$H_{int} = \int d\sigma \left(\lambda_1 J_1(\sigma) J_2(\sigma) + \lambda_2 \sum_n \alpha_n^\mu \tilde{\alpha}_{-n,\mu} \right)$$

[45.0]

where λ_1 and λ_2 are coupling constants, and $J_1(\sigma)$ and $J_2(\sigma)$ represent the current densities along the strings.

4.2.3. Topological Quantum Field Theory (TQFT):
- TQFT provides a framework for understanding the entanglement of strings based on their topological properties. In TQFT, the entanglement entropy can be computed for regions of space containing topologically equivalent strings.
- Entanglement Entropy: The entanglement entropy S_A of a region A containing a segment of a string can be expressed in TQFT as

$$S_A = -\text{Tr}(\rho_A \log \rho_A)$$

[46.0]

where ρ_A is the reduced density matrix for the region A.

Practical Implementation

1. Experimental Realization of Topological Strings:
   - Creating and controlling strings with specific topological configurations in a laboratory setting can be challenging. However, advances in materials science and quantum technology provide pathways to simulate such conditions.
   - Case: Topological insulators and superconducting circuits can be engineered to mimic the properties of open and closed strings, allowing experimental study of their interactions and entanglement.

2. Measurement Techniques:
   - Detecting the vibrational modes and interactions of topological strings requires sophisticated measurement tools. Techniques such as scanning tunneling microscopy (STM) and quantum state tomography can be employed.
   - Case: STM can visualize the edge states of topological insulators, providing insights into the system's vibrational modes and potential entanglement properties.

Strings' topological properties, such as whether they are open or closed, play a crucial role in their interactions and potential for entanglement. Strings with similar topological configurations can synchronize their vibrational modes and interact through specific channels, leading to entangled states. The use of advanced experimental techniques opens up exciting possibilities for discovery in this field. By exploring these topological characteristics and employing these techniques, researchers can gain deeper insights into the mechanisms of quantum entanglement in string theory.

5.0. Dimensional Embedding:

Concept

In string theory, strings are not confined to the familiar three spatial dimensions but instead exist in higher-dimensional spaces. These additional dimensions, which are compactified and not directly observable, play a crucial role in determining the properties and interactions of strings. The way strings are embedded in these higher-dimensional spaces can significantly affect their behavior and potential for entanglement. Shared dimensional embeddings can create a basis for entanglement by aligning the vibrational modes and other properties of strings.

Importance in Quantum Entanglement

The dimensional embedding of strings determines the allowed vibrational modes, interactions, and other physical properties. When strings are embedded in similar or compatible higher-dimensional spaces, their quantum states can become correlated, leading to entanglement. This entanglement is a direct consequence of the geometric and topological properties of the extra dimensions in which the strings reside.

Mathematical Representation

5.1. Higher-Dimensional Spaces:
- Strings exist in D-dimensional spacetime where $D > 4$ (*typically $D =$ or $D = 1$) in various string theory models*). The extra dimensions are usually compacted on small scales.
- The coordinates of a string in D-dimensional spacetime can be expressed as

$$X^\mu(\tau, \sigma), \quad \mu = 0, 1, 2, \ldots, D-1$$

[47.0]

where τ and σ are the worldsheet coordinates.

5.2. Compactification:
- The extra dimensions are compactified on small scales, often described by a compact manifold M. A common choice is the six-dimensional Calabi-Yau manifold in 10-dimensional superstring theory.
- The compactification can be mathematically described using a metric g_{IJ} on the compact manifold:

$$ds^2 = g_{ij} dy^i dy^j$$

[48.0]

where $i, j = 1, \ldots, 6$ are the coordinates of the compact dimensions.

5.3. Vibrational Modes and Embeddings:
- The vibrational modes of strings depend on their embedding in the compactified dimensions. For Case, a string winding around a compact dimension y^i will have quantized vibrational modes determined by the geometry of the compactification.
- The energy of a mode n in a compact dimension with radius R_i is given by

$$E_n = \frac{n}{R_i}$$

[49.0]

5.4. Wavefunction Representation:
- The quantum state of a string embedded in higher dimensions can be represented by a wavefunction $\psi(X^\mu)$ that incorporates the extra-dimensional coordinates.
- For two entangled strings, the combined wavefunction might be:

$$\Psi(X_1^\mu, X_2^\mu) = \sum_{m,n} c_{mn} \psi_m(X_1^\mu) \phi_n(X_2^\mu)$$

[50.0]

where X_1^μ and X_2^μ are the coordinates of the two strings, and ψ_m and ϕ_n are the wavefunctions corresponding to the vibrational modes influenced by the dimensional embedding.

Implications for Entanglement

5.6. Geometric and Topological Alignment:
- When strings share similar embeddings in the compactified dimensions, their vibrational modes and interactions become correlated. This alignment can lead to entanglement.
- Case: Two strings embedded in a Calabi-Yau manifold with the same homology cycles can exhibit correlated vibrational modes, leading to entanglement.

5.7. Overlap Integrals:
- The degree of overlap between the wavefunctions of two strings in the extra dimensions determines the strength of their entanglement. The overlap integral can be expressed as

$$I_{mn} = \int d^6 y \, \psi_m^*(y) \phi_n(y)$$

[51.0]

where y represents the coordinates of the compact dimensions.

5.8. Dimensional Reduction and Effective Theory:
- The effects of the extra dimensions can be seen in the lower-dimensional effective theory obtained by dimensional reduction. The resulting effective field theory incorporates the influence of the higher-dimensional embedding.
- Effective Action: The effective action in the lower-dimensional theory can include terms that reflect the correlations induced by the higher-dimensional embeddings

$$S_{\text{eff}} = \int d^4x \sqrt{-g}\left(\frac{1}{2}g^{\mu\nu}\partial_\mu\phi\partial_\nu\phi - V_{\text{eff}}(\phi)\right)$$

[52.0]

where $V_{eff}(\phi)$ includes contributions from the compactified dimensions.

Practical Implementation

5.0.1. Simulating Higher-Dimensional Embeddings:
- While directly accessing higher-dimensional spaces is beyond current experimental capabilities, analogous systems can be created to simulate these effects. Quantum simulators and engineered materials can mimic the properties of higher-dimensional embeddings.
- Case: Optical lattices and superconducting qubit arrays can be designed to exhibit behaviors analogous to strings embedded in higher dimensions.

5.0.2. Measurement Techniques:
- Advanced measurement techniques, such as interferometry and quantum state tomography, are required to detect the subtle effects of higher-dimensional embeddings on entanglement.
- Case: Quantum state tomography can reconstruct the quantum state of strings or analogous systems, revealing the correlations induced by their higher-dimensional embeddings.

Dimensional embedding plays a crucial role in the behavior and potential for entanglement of strings. Strings embedded in similar or compatible higher-dimensional spaces can exhibit correlated vibrational modes and interactions, leading to entanglement. By understanding the geometric and topological properties of these embeddings, researchers can gain deeper insights into the mechanisms of quantum entanglement in string theory. The role of advanced experimental techniques and simulations is pivotal in exploring these effects, providing a bridge between theoretical predictions and observable phenomena.

6. Gauge Symmetries:

Concept

Gauge symmetries are fundamental symmetries in field theories that describe the interactions of elementary particles. In string theory, gauge symmetries play a crucial role in determining how strings interact with each other. These symmetries govern the allowable transformations of the fields without altering the physical content of the theory. When strings share the same gauge symmetries, their interactions can lead to correlated behaviors that are indicative of entanglement.

Importance in Quantum Entanglement

Shared gauge symmetries between strings can facilitate entanglement by ensuring that the strings interact in a synchronized and correlated manner. These symmetries can lead to the alignment of quantum states, allowing for the entanglement of particles even across large distances. The correlated behaviors arising from shared gauge symmetries can be mathematically described and measured, providing a clear basis for understanding entanglement.

Mathematical Representation

1. Gauge Fields and Symmetries:
   - Gauge symmetries are described by gauge fields A_μ^a, where a indexes the components of the field in the gauge group and μ indexes spacetime coordinates.
   - The gauge transformations are given by

$$A_\mu \to A'_\mu = U A_\mu U^{-1} + \frac{i}{g}(\partial_\mu U) U^{-1}$$

[53.0]

where U is an element of the gauge group and g is the coupling constant.

2. Field Strength Tensor:
   - The dynamics of gauge fields are encoded in the field strength tensor $F_{\mu\nu}^a$ defined as

$$F_{\mu\nu}^a = \partial_\mu A_\nu^a - \partial_\nu A_\mu^a + g f^{abc} A_\mu^b A_\nu^c$$

[54.0]

where f^{abc} are the structure constants of the gauge group.

3. Yang-Mills Lagrangian:
   - The Lagrangian for a gauge field is given by the Yang-Mills Lagrangian:

$$\mathcal{L}_{YM} = -\frac{1}{4} F^a_{\mu\nu} F^{\mu\nu a}$$

[55.0]

4. Wavefunction and Shared Gauge Symmetries:
   - The quantum state of a string in a gauge theory can be represented by a wavefunction $\psi(A_\mu)$. For two entangled strings, the combined wavefunction might be

$$\Psi(A^1_\mu, A^2_\mu) = \sum_{m,n} c_{mn} \psi_m(A^1_\mu) \phi_n(A^2_\mu)$$

[56.0]

where A^1_μ and A^2_μ are the gauge fields associated with the two strings.

5. Gauge Invariance and Correlation:
   - The entangled state must be gauge invariant, meaning that the physical observables are unchanged under gauge transformations. This can be enforced by ensuring that the wavefunction is invariant under simultaneous gauge transformations.

$$\Psi(A^1_\mu, A^2_\mu) = \Psi(U_1 A^1_\mu U_1^{-1} + \frac{i}{g}(\partial_\mu U_1) U_1^{-1}, U_2 A^2_\mu U_2^{-1} + \frac{i}{g}(\partial_\mu U_2) U_2^{-1})$$

[57.0]

Implications for Entanglement

1. Correlated Interactions:
   - Shared gauge symmetries ensure that the strings interact through the same gauge fields, leading to correlated quantum states. This correlation is a key feature of entanglement.
   - Case: Two strings interacting through a common SU(3) gauge field will have their quantum states correlated due to the shared symmetry structure.

2. Gauge Boson Exchange:
   - The exchange of gauge bosons (*force carriers*) between strings can lead to entanglement. The gauge bosons mediate interactions that are consistent with the shared gauge symmetries.
   - Interaction Hamiltonian: The interaction Hamiltonian for gauge boson exchange can be written as:

$$H_{int} = g \int d^4x \, J^\mu A_\mu$$

[58.0]

where J^μ is the current density and A_μ is the gauge field.

3. Topological Effects:
- Gauge symmetries can also give rise to topological effects, such as flux tubes or monopoles, which can further enhance entanglement. These topological features are protected by the gauge symmetry and contribute to the robustness of the entangled state.

Practical Implementation

1. Experimental Realization of Gauge Symmetries:
- Realizing shared gauge symmetries in a laboratory setting involves creating systems where the gauge fields can be precisely controlled. This can be achieved using superconducting circuits, cold atom systems, or photonic crystals.
- Case: Superconducting qubits can be engineered to have interactions mediated by effective gauge fields, allowing the study of gauge symmetry-induced entanglement.

2. Measurement Techniques:
- Detecting the correlated behaviors indicative of entanglement requires precise measurement of the gauge fields and the quantum states of the strings. Techniques such as quantum state tomography and interference measurements can be used.
- Case: Quantum state tomography can reconstruct the quantum state of the system, revealing the correlations arising from shared gauge symmetries.

Gauge symmetries, a fundamental aspect in the interactions and potential entanglement of strings, hold the key to a world of discoveries. Shared gauge symmetries ensure that strings interact in a correlated manner, leading to entangled quantum states. By understanding and controlling these symmetries, researchers can unlock new mechanisms of quantum entanglement, potentially revolutionizing our understanding of the fundamental nature of quantum mechanics. Advanced experimental techniques and theoretical models provide a pathway to explore these effects in detail, sparking excitement about the possibilities that lie ahead.

7. Coupling Constants:

Concept

In string theory, coupling constants determine the strength of interactions between strings. These constants are fundamental parameters that influence how strings interact, exchange energy, and correlate their quantum states. When strings have similar or harmonically related coupling constants, their interactions can become synchronized, leading to correlated quantum states and potential entanglement. The coupling constants can be thought of as the '*interaction coefficients*' that dictate the intensity and nature of the forces between strings.

Importance in Quantum Entanglement

Similar coupling constants between strings can lead to correlated behaviors, as the interaction strengths determine the dynamics of the system. These correlations are a crucial aspect of entanglement, where the quantum states of particles become interdependent. By examining the coupling constants, we can predict and understand how strings will interact and become entangled.

Mathematical Representation

1. Interaction Strength and Coupling Constants:
   - The interaction strength between two strings is determined by the coupling constant g. The general form of the interaction Hamiltonian can be expressed as

$$H_{int} = g \int d\sigma \, J(\sigma)$$

[59.0]

where $J(\sigma)$ is the current density along the string and σ is the spatial coordinate on the string.

2. Effective Coupling in Higher Dimensions:
   - In higher-dimensional theories, the effective coupling constant can be influenced by the geometry of the compactified dimensions. For Case, in a compactification scheme, the effective coupling g_{eff} might be related to the original coupling g and the volume V of the compact dimensions:

$$g_{\text{eff}} = \frac{g}{\sqrt{V}}$$

[60.0]

3. Correlation of Quantum States:
   - For two strings with coupling constants g_1 and g_2, their quantum states can become correlated if $g_1 \approx g_2$. The combined wavefunction of the entangled strings can be expressed as

$$\Psi(g_1, g_2) = \sum_{m,n} c_{mn} \psi_m(g_1) \phi_n(g_2)$$

[61.0]

where ψ_m and ϕ_n are the wavefunctions corresponding to the vibrational modes of the strings and c_{mn} are coefficients representing the degree of entanglement.

Implications for Entanglement

1. Interaction Dynamics:
   - Strings with similar coupling constants will interact with similar strengths, leading to synchronized dynamics. This synchronization can enhance the probability of entanglement as the quantum states of the strings become interdependent.
   - Case: Two strings in a system where $g_1 \approx g_2$ will exchange energy and information more efficiently, leading to a higher likelihood of becoming entangled.

2. Resonance Effects:
   - When the coupling constants are harmonically related, resonance effects can occur, further strengthening the entanglement. These resonances can amplify specific vibrational modes, aligning the quantum states of the strings.
   - Resonance Condition: If $g_1 = kg_2$ (where k) is a rational number), the resonance can enhance the entanglement between the strings.

3. Non-linear Interactions:
   - Non-linear interactions in the string dynamics, influenced by the coupling constants, can lead to complex entangled states. These interactions can be described by higher-order terms in the interaction Hamiltonian.

$$H_{int} = g \int d\sigma \left(J(\sigma) + \lambda J^2(\sigma) + \mu J^3(\sigma) \right)$$

[62.0]

where λ and μ are higher-order coupling constants.

Practical Implementation

1. Experimental Realization of Coupling Constants:
   - Achieving precise control over the coupling constants in experimental setups is crucial for studying entanglement. This can be done using engineered materials, such as superconducting qubits or cold atom systems, where the interaction strengths can be finely tuned.
   - Case: Superconducting qubits can be designed with tunable Josephson junctions, allowing precise control over the coupling constants.

2. Measurement Techniques:
- Measuring the coupling constants and the resulting quantum states requires advanced techniques like quantum state tomography, spectroscopic analysis, and interferometry.
- Case: Quantum state tomography can reconstruct the entangled state of the system, providing detailed information about the correlation induced by similar coupling constants.

Coupling constants, as fundamental parameters, play a crucial role in determining the interaction strengths between strings in string theory. When these constants are similar or harmonically related, they can lead to correlated quantum states, thereby facilitating entanglement. The understanding and control of these constants are key to exploring the mechanisms of quantum entanglement. This exploration is made possible by the use of advanced experimental techniques and theoretical models, which allow researchers to investigate the role of coupling constants in the creation and maintenance of entangled states.

Experimental Proposals

Particle Generation and Synchronization

A1. Particle Accelerators:

Concept: High-energy particle accelerators are instrumental in generating particles with specific string characteristics. These accelerators can create the extreme conditions necessary to probe the fundamental properties of particles and strings, allowing researchers to explore vibrational modes, gauge symmetries, and other string theory predictions.

Detailed Proposal

Objective:
- To generate particles with specific string characteristics and investigate their properties to test the hypothesis that shared string characteristics lead to quantum entanglement.

Methodology:
- High-Energy Collisions: Utilize high-energy collisions in particle accelerators to produce particles and analyze their string-like properties.
- Synchronization Techniques: Implement advanced synchronization techniques to ensure that the generated particles can be studied in correlated states.

Steps Involved:

1. High-Energy Collisions:
   - Particle Accelerators: To achieve the necessary energy levels, facilities like the Large Hadron Collider (LHC) or future colliders like the International Linear Collider (ILC) will be used.
   - Collision Dynamics: Particles are accelerated to near-light speeds and then collide, producing a range of secondary particles.
   - String Characteristic: Analyze the properties of these particles to identify those that exhibit string characteristics, such as specific vibrational modes or gauge symmetries.

Mathematical Framework:
   - Energy and Momentum Conservation:

$$E_{\text{initial}} = E_{\text{final}}, \quad \mathbf{p}_{\text{initial}} = \mathbf{p}_{\text{final}}$$

[63.0]

where E represents energy and $\mathbf{p}$ represents momentum.

   - String Vibrational Modes:

$$E_n = \frac{n}{R_i}, \quad \omega_n = \frac{n\pi}{L}\sqrt{\frac{T}{\mu}}$$

[63.1]

where n is the vibrational mode number, R_i is the compactification radius, T is the string tension, and μ is the linear mass density.

2. Synchronization Techniques:
- Synchronized Lasers: Use synchronized laser pulses to initiate and control particle interactions. Lasers can be tuned to specific frequencies to match the vibrational modes of the particles.
- Atomic Clocks: Employ atomic clocks to achieve precise temporal synchronization of particle states. This ensures that the phases of the quantum states are aligned, facilitating entanglement.

Innovative Approaches:

1. Hybrid Acceleration Methods:
- Combine traditional particle acceleration with laser-plasma acceleration techniques to achieve higher energy densities and more controlled particle generation. Laser-plasma acceleration uses intense laser pulses to create plasma waves, which can accelerate particles to high energies over shorter distances.

Mathematical Expression:

$$E_{\text{plasma}} = \frac{eE_{\text{wave}}\lambda}{2\pi}$$

[64.0]

where E_{plasma} is the energy of the plasma wave, e is the electron charge, E_{wave} is the electric field of the wave, and λ is the wavelength of the plasma wave.

B1. Quantum Simulation:

- Use quantum simulators to mimic the conditions in high-energy particle accelerators. Quantum simulators can create and manipulate states that resemble those of strings, providing a more accessible way to study entanglement and other properties.
- Simulation Models: Implement models that replicate the Hamiltonian dynamics of string interactions

$$H = \sum_i \left(\frac{p_i^2}{2m} + V(x_i) \right) + \sum_{i<j} U(x_i, x_j)$$

[65.0]

where p_i and x_i are the momentum and position of particle i, $V(x_i)$ is the potential energy, and $U(x_i, x_j)$ is the interaction potential.

3. Advanced Detection and Measurement:

- Develop new detection methods that enhance the resolution and sensitivity of particle measurements. Techniques such as superconducting nanowire single-photon detectors (SNSPDs) and advanced interferometry can provide precise measurements of particle properties and interactions.
- Detection Metrics: Measure key parameters such as energy levels, phase coherence, and interaction cross-sections:

$$\sigma = \frac{N_{\text{events}}}{L_{\text{int}} \cdot \epsilon}$$

[66.0]

where σ is the cross-section, N_{events} is the number of observed events, L_{int} is the integrated luminosity, and ϵ is the detection efficiency.

High-energy particle accelerators are powerful tools that allow us to generate particles with specific string characteristics. What's truly exciting is our innovative approach. By combining advanced synchronization techniques, cutting-edge acceleration methods, and state-of-the-art detection technologies, we're able to study the properties of these particles. This unique approach not only enhances our understanding of string theory but also pushes the boundaries of experimental physics, sparking new possibilities and discoveries.

B. Quantum Simulators:

Concept

Accurately measuring the coupling constants and the resulting quantum states of particles generated in high-energy experiments is crucial for validating theories and hypotheses in quantum mechanics and string theory. Advanced measurement techniques such as quantum state tomography, spectroscopic analysis, and interferometry play essential roles in reconstructing the entangled states and understanding the correlations induced by similar coupling constants.

Importance in Quantum Entanglement

These measurement techniques provide the tools necessary to observe and quantify the quantum states of particles, ensuring that the theoretical predictions match experimental outcomes. By employing precise and sophisticated methods, researchers can detect subtle features of entanglement and the effects of coupling constants on particle interactions.

Measurement Techniques

1. Quantum State Tomography:
    - Concept: Quantum state tomography is a process of reconstructing the quantum state of a system by measuring various properties and using the data to infer the state. It involves performing a series of measurements on a quantum system and using statistical methods to reconstruct the state.

Mathematical Framework:

- The density matrix ρ of a quantum state can be reconstructed from measurement data

$$\rho = \sum_{i,j} \rho_{ij} |i\rangle \langle j|$$

- To reconstruct ρ, measurements in different bases are required. The probability of measuring outcome m in basis b is given by

$$P(m|b) = \text{Tr}(\rho M_{m|b})$$

where $M_{m|b}$ is the measurement operator.

- Case: Quantum state tomography can reconstruct a system's entangled state, providing detailed information about the correlation induced by similar coupling constants. For instance, in a two-qubit system, measurements in the Pauli bases can be used to determine the state fully.

2. Spectroscopic Analysis:
- Concept: Spectroscopy involves studying the interaction between matter and electromagnetic radiation to infer the properties of particles. It is used to measure energy levels, transition frequencies, and other properties that provide insights into the coupling constants and the structure of quantum states.

Mathematical Framework:
- The energy levels E_n of a system can be related to the observed spectral lines:

$$E_n = h\nu_n$$

[68.0]

where h is Planck's constant and ν_n is the frequency of the spectral line.
- The intensity of spectral lines can provide information about the population of energy states and transition probabilities.
- Case: In particle accelerators, spectroscopic analysis can be used to measure the energy levels and transition frequencies of generated particles, helping to determine the coupling constants.

3. Interferometry:
- Concept: Interferometry uses the principle of interference to measure small distances, displacements, and changes in optical paths. It is highly sensitive and can detect phase shifts and the coherence properties of quantum states.

Mathematical Framework:
- The phase difference Δϕ between two paths in an interferometer is related to the path length difference ΔL:

$$\Delta\phi = \frac{2\pi}{\lambda}\Delta L$$

[69.0]

where λ is the wavelength of the light used.
- The interference pattern observed is given by:

$$I(\Delta\phi) = I_0\left(1 + \cos(\Delta\phi)\right)$$

[69.1]

where I_0 is the maximum intensity.
- Case: Interferometry can measure the coherence length and phase shifts in quantum states, providing precise information about the entangled states and their properties.

Innovative Approaches

1. Integrated Quantum Sensors:
   - Develop integrated quantum sensors that combine elements of quantum state tomography, spectroscopy, and interferometry in a single device. These sensors can provide comprehensive, high-precision measurements of quantum states.
   - Case: A hybrid quantum sensor using NV centers in diamond can perform quantum state tomography and spectroscopic analysis simultaneously, providing detailed information about the quantum state and its environment.

2. Machine Learning in Quantum Measurements:
   - Machine learning algorithms analyze measurement data and improve the accuracy of quantum state reconstruction. These algorithms can handle large datasets and identify patterns that traditional methods might miss.

Mathematical Framework:
   - Machine learning models can be trained to predict the density matrix ρ from measurement outcomes:

$$\hat{\rho} = f(\{P(m|b)\}; \theta)$$

[70.0]

   where f is a machine learning model parameterized by θ.
   - Case: Neural networks can be trained to reconstruct quantum states from tomography data, providing faster and more accurate state determination.

3. Adaptive Measurement Techniques:
   - Implement adaptive measurement techniques that dynamically adjust the measurement parameters based on real-time data. This approach optimizes the measurement process and improves the quality of the reconstructed quantum states.
   - Case: Adaptive quantum state tomography uses feedback from initial measurements to refine the subsequent measurement settings, reducing the number of measurements needed for accurate state reconstruction.

Advanced measurement techniques are indispensable for accurately determining the coupling constants and quantum states in high-energy physics experiments. Quantum state tomography, spectroscopic analysis, and interferometry provide the tools needed to reconstruct and analyze entangled states. However, the true power lies in integrating these techniques with innovative approaches such as quantum sensors, machine learning, and adaptive measurements. This integration allows us to achieve unprecedented precision in studying quantum entanglement and string theory predictions, opening up new avenues for research and discovery.

C1. Temporal Synchronization:

Concept

Temporal synchronization refers to the precise alignment of the timing of particles' quantum states. This synchronization is crucial for maintaining coherence and achieving entanglement between particles. By utilizing advanced technologies such as synchronized lasers and atomic clocks, researchers can ensure that the quantum states of particles are temporally aligned, facilitating robust quantum entanglement.

Importance in Quantum Entanglement

Temporal synchronization ensures that particles' quantum phases are aligned, which is essential for maintaining coherence and enabling entanglement. Without precise synchronization, even small temporal discrepancies can lead to decoherence, which destroys the entangled states.

Measurement Techniques

1. Synchronized Lasers:
    - Concept: Synchronized lasers can produce pulses with precise timing, ensuring that particles interact at the same moment. This synchronization helps in maintaining the coherence of the quantum

Mathematical Framework:

- The phase of the laser pulses can be described by:

$$E(t) = E_0 e^{i(\omega t + \phi(t))}$$

[71.0]

where E_0 is the amplitude, ω is the angular frequency and $\phi(t)$ is the phase as a function of time.

- Synchronizing the phases of multiple lasers involves matching $\phi(t)$ across all laser sources:

$$\phi_1(t) = \phi_2(t) = \phi_3(t) = \ldots$$

[71.1]

- Case: In quantum communication, synchronized laser pulses can be used to ensure that entangled photons are emitted and detected simultaneously, preserving their entangled states.

2. Atomic Clocks:
- Concept: Atomic clocks provide the most precise timekeeping, using the vibrations of atoms (such as cesium or rubidium) as a frequency standard. These clocks ensure that the temporal alignment of quantum states is maintained over long durations.

Mathematical Framework:
- The frequency of an atomic clock is given by:

$$\nu = \frac{\Delta E}{h}$$

[72.0]

where ΔE is the energy difference between two atomic states and h is Planck's constant.

- The precision of atomic clocks can be characterized by their stability, often measured as the Allan deviation,

$$\sigma_y(\tau) = \sqrt{\frac{1}{2}\left\langle \left(\frac{\Delta y_{n+1} - \Delta y_n}{\tau}\right)^2 \right\rangle}$$

where Δy is the fractional frequency deviation and τ is the averaging time.
- Case: In quantum experiments, atomic clocks can be used to synchronize the timing of particle interactions, ensuring that entangled states remain coherent over extended periods.

Innovative Approaches

1. Optical Lattice Clocks:
   - Develop and utilize optical lattice clocks, which trap atoms in a lattice formed by intersecting laser beams. These clocks offer even greater precision than traditional atomic clocks and can be used to synchronize quantum experiments with extreme accuracy.

Mathematical Framework:
   - The trapping potential in an optical lattice can be described as:

$$V(x,t) = V_0 \sin^2(kx - \omega t)$$

where V_0 is the depth of the potential, k is the wavevector, and ω is the angular frequency of the lattice beams.
   - Case: Optical lattice clocks can synchronize the quantum states of particles in different locations, enhancing the robustness of entangled states used in quantum networks.

2. Quantum Dot Synchronization:
   - Use quantum dots, which are nanoscale semiconductor particles, as sources of single photons or entangled photon pairs. By synchronizing the emission times of quantum dots using pulsed lasers, researchers can create highly coherent entangled states.

Mathematical Framework:
- The timing of the excitation laser pulse can control the emission time of a quantum dot:

$$t_{\text{emission}} = t_{\text{pulse}} + \tau_{\text{excitation}}$$

[74.0]

where t_{pulse} is the time of the laser pulse and $\tau_{excitation}$ is the excitation time of the quantum dot.

- Case: Quantum dots synchronized with pulsed lasers can be used to generate entangled photons on demand and are suitable for quantum communication and computation applications.

3. Entanglement Swapping with Time-Bin Encoding:
- Implement entanglement swapping protocols using time-bin encoding, where the entanglement is encoded at the time of arrival of photons. This method requires precise temporal synchronization to ensure that the entangled states are correctly correlated.

Mathematical Framework:

- The time-bin entangled state can be described as

$$|\psi\rangle = \frac{1}{\sqrt{2}} \left(|t_1\rangle_A |t_1\rangle_B + |t_2\rangle_A |t_2\rangle_B \right)$$

[75.0]

where $|t_1\rangle$ and $|t_2\rangle$ represent different time bins.

- Case: Time-bin entanglement swapping can be used in quantum repeaters to extend the range of quantum communication networks, with synchronized detection ensuring the fidelity of the entangled states.

Temporal synchronization is vital for maintaining coherence and enabling entanglement in quantum systems. By using synchronized lasers and atomic clocks, researchers can achieve precise temporal alignment of quantum states. Innovative approaches such as optical lattice clocks, quantum dot synchronization, and entanglement swapping with time-bin encoding further enhance the ability to create and manipulate entangled states with high precision.

Measurement and Correlation Analysis

A2. Quantum Sensors:

Concept

Quantum sensors are advanced devices that exploit quantum mechanical properties to measure physical quantities with exceptional precision. They are essential for detecting and analyzing the properties of particles in quantum experiments. These sensors can measure various properties, including position, momentum, energy, and quantum states, enabling detailed correlation analysis between entangled particles.

Importance in Quantum Entanglement

Advanced quantum sensors are crucial for validating the presence and quality of entanglement. By accurately measuring particles' properties, these sensors help establish the correlations predicted by quantum mechanics.
This information is vital for experiments that test the fundamental principles of quantum theory and for applications in quantum computing and communication.

Measurement Techniques

1. Quantum Sensors:
    - Concept: Quantum sensors leverage the principles of quantum mechanics, such as superposition and entanglement, to achieve high sensitivity and precision. These sensors are used to measure various properties of particles, including their quantum states, with minimal disturbance.

- *Types of Quantum Sensors:*
  - Superconducting Quantum Interference Devices (SQUIDs): Measure magnetic fields with extreme precision.
  - Nitrogen-Vacancy (NV) Centers in Diamond: Detect magnetic and electric fields at the atomic scale.
  - Atomic Interferometers: Measure gravitational forces and accelerations with high accuracy.

Mathematical Framework:
- Measurement Operators:
  - The measurement process in quantum mechanics is described by operators that act on the quantum state. For a property A, the measurement operator is $\hat{A}$.
    - The expectation value of the measurement is given by:

$$\langle \hat{A} \rangle = \langle \psi | \hat{A} | \psi \rangle$$

[76.0]

where $|\psi\rangle$ is the quantum state of the system.

Measurement Precision:

- The Heisenberg uncertainty principle often limits the precision of a measurement:

$$\Delta A \cdot \Delta B \geq \frac{\hbar}{2}$$

[76.1]

where ΔA and ΔB are the standard deviations of two complementary properties A and B.

2. Correlation Analysis:

- Concept: Correlation analysis involves studying the statistical relationships between the properties of entangled particles. By measuring these correlations, researchers can verify the presence of entanglement and quantify its strength.

Mathematical Framework:

- Correlation Function:

$$E(\hat{a}, \hat{b}) = \langle \psi | \hat{a} \otimes \hat{b} | \psi \rangle$$

[77.0]

where $\hat{a}$ and $\hat{b}$ are measurement operators for the two entangled particles.

- Bell's Inequality:

$$S = |E(\hat{a}, \hat{b}) - E(\hat{a}, \hat{b}') + E(\hat{a}', \hat{b}) + E(\hat{a}', \hat{b}')| \leq 2$$

[77.1]

- Violation of Bell's inequality indicates the presence of quantum entanglement.

Innovative Approaches

1. Integrated Quantum Sensor Networks:
   - Develop networks of integrated quantum sensors that can communicate and share data in real-time. These networks can provide comprehensive measurements of entangled states across different locations.
   - Case: A network of NV centers in diamond distributed across a large area can measure the magnetic properties of entangled particles with high precision, allowing for spatial correlation analysis.

2. Hybrid Quantum-Classical Sensors:
   - Combine quantum sensors with classical sensors to enhance measurement capabilities. These hybrid systems can leverage the high sensitivity of quantum sensors and the robustness of classical sensors.

Mathematical Framework:
   - The combined measurement operator for a hybrid sensor system can be expressed as

$$\hat{A}_{\text{hybrid}} = \alpha \hat{A}_{\text{quantum}} + \beta \hat{A}_{\text{classical}}$$

[78.0]

where α and β are weighting factors.
   - Case: A hybrid sensor system using SQUIDs for magnetic field measurements and traditional Hall effect sensors can provide comprehensive data on the magnetic properties of entangled states.

3. Machine Learning for Correlation Analysis:
   - Utilize machine learning algorithms to analyze measurement data and identify correlations that indicate entanglement. These algorithms can handle large datasets and uncover patterns that may take time to be apparent.

Mathematical Framework:

- Machine learning models can be trained to recognize entangled states based on measurement outcomes:

$$\hat{\rho} = f(\{E(\hat{a}, \hat{b})\}; \theta)$$

where f is a machine learning model parameterized by θ.

- Case: Neural networks can analyze data from quantum sensors to classify the degree of entanglement between particles, providing insights into the strength and nature of the correlations.

Our research on advanced quantum sensors and correlation analysis techniques holds immense potential for measuring and understanding the properties of entangled particles. By utilizing integrated quantum sensor networks, hybrid quantum-classical sensors, and machine learning algorithms, we are on the brink of achieving unprecedented precision in detecting and analyzing entangled states. These innovations are set to advance our understanding of quantum mechanics significantly and catalyze the development of practical quantum technologies, promising a future of limitless possibilities.

B2. Correlation Studies:

Concept

Correlation studies are fundamental to determining the presence of quantum entanglement between particles. By analyzing the correlations between particles' properties, researchers can identify entangled states and quantify the degree of entanglement. These studies involve statistical analysis of measurement outcomes to reveal the non-classical correlations characteristic of entanglement.

Importance in Quantum Entanglement

The presence of entanglement manifests as strong correlations between the measurement outcomes of entangled particles, which cannot be explained by classical physics. By performing detailed correlation studies, researchers can verify entanglement and investigate its properties. These studies are essential for validating quantum theories and developing practical quantum technologies.

Measurement Techniques

1. Bell Test Experiments:
   - Concept: Bell test experiments are designed to measure the correlations between entangled particles and test quantum mechanics' predictions against classical theories. They involve measuring particles' properties along different axes and comparing the results to Bell's inequalities.

Mathematical Framework:
   - The correlation function for measurements along axes $\hat{a}$ and $\hat{b}$ is given by:

$$E(\hat{a}, \hat{b}) = \langle \psi | \hat{\sigma} \cdot \hat{a} \otimes \hat{\sigma} \cdot \hat{b} | \psi \rangle$$

[80.0]

where $\hat{\sigma}$ are the Pauli matrices and $|\psi\rangle$ is the entangled state.

   - Bell's inequality for two particles is expressed as

$$S = |E(\hat{a}, \hat{b}) - E(\hat{a}, \hat{b}') + E(\hat{a}', \hat{b}) + E(\hat{a}', \hat{b}')| \leq 2$$

[80.1]

Violation of this inequality indicates the presence of entanglement.

2. Quantum State Tomography:

- Concept: Quantum state tomography involves reconstructing the quantum state of a system based on measurements in multiple bases. This technique provides detailed information about the correlations between particles' properties.

Mathematical Framework:

- The density matrix ρ of a quantum state can be reconstructed using measurements in different bases:

$$\rho = \sum_{i,j} \rho_{ij} |i\rangle\langle j|$$

[81.0]

- The probabilities of measurement outcomes are used to estimate the elements of the density matrix:

$$P(m|b) = \text{Tr}(\rho M_{m|b})$$

[81.1]

where $M_{m|b}$ are the measurement operators for basis b and outcome m.

3. Entanglement Witnesses:
- Concept: Entanglement witnesses are operators that can detect entanglement by providing a value that indicates whether a given state is entangled.

Mathematical Framework:
- An entanglement witness w is an observable for which the expectation value is negative for entangled states and non-negative for separable states:

$$\langle W \rangle = \text{Tr}(W\rho)$$

[82.0]

- For Case, the witness for a two-qubit state might be:

$$W = I - |\phi^+\rangle\langle\phi^+|$$

[82.1]

where $|\phi^+\rangle$ is a maximally entangled Bell state and I is the identity operator.

Innovative Approaches

1. Machine Learning for Correlation Analysis:
- Machine learning algorithms analyze large datasets of measurement outcomes and identify correlations indicative of entanglement. These algorithms can uncover patterns that may take time to become apparent using traditional methods.

Mathematical Framework:

- Machine learning models can be trained to predict the presence of entanglement from measurement data:

$$\hat{E} = f(\{P(m|b)\}; \theta)$$

[83.0]

where f is a machine learning model parameterized by θ, and $P(m|b)$ are the measurement probabilities.

- Case: Neural networks can classify measurement data to determine the presence and strength of entanglement, providing a powerful tool for analyzing complex quantum systems.

2. Real-Time Correlation Analysis with Quantum Sensors:

- Develop quantum sensors capable of performing real-time correlation analysis. These sensors can continuously monitor and analyze the properties of entangled particles, providing immediate feedback on the presence of entanglement.

Mathematical Framework:

- The real-time correlation function can be updated dynamically:

$$E_{\text{real-time}}(\hat{a}, \hat{b}, t) = \frac{1}{N(t)} \sum_{i=1}^{N(t)} \sigma_i(\hat{a})\sigma_i(\hat{b})$$

[84.0]

where $N(t)$ is the number of measurements performed up to time t.

- Case: Real-time quantum sensors can be used in quantum communication networks to monitor entanglement between nodes, ensuring robust and reliable quantum links.

3. Hybrid Quantum-Classical Systems for Enhanced Correlation Detection:
- Combine quantum and classical systems to enhance the detection and analysis of correlations. Classical systems can preprocess data and identify potential correlations, which are then confirmed and refined by quantum systems.

Mathematical Framework:
- The hybrid system can use classical preprocessing to filter measurement data:

$$\hat{P}_{\text{classical}} = g(\{P(m|b)\})$$

[85.0]

where g is a classical preprocessing function. A quantum system then analyzes the filtered data:

$$E_{\text{hybrid}} = \sum_i f(\hat{P}_{\text{classical},i})\hat{Q}_i$$

[85.1]

where $\hat{Q}_i$ are quantum operators.
- Case: Hybrid systems can be used in quantum computing to preprocess large datasets, reducing the computational load on quantum processors and improving the efficiency of correlation analysis.

Correlation studies are essential for identifying and quantifying quantum entanglement. By analyzing the correlations between particles' properties, researchers can validate entanglement and explore its properties. Advanced techniques such as Bell test experiments, quantum state tomography, and entanglement witnesses provide robust methods for correlation analysis. But what's truly exciting are the innovative approaches that lie ahead. Imagine the potential of machine learning, real-time quantum sensors, and hybrid quantum-classical systems, offering enhanced capabilities for detecting and understanding entanglement in complex quantum systems. These are the tools that will shape the future of our field.

Advanced Experimental Setups

A3. Quantum Dot Arrays:

Concept:

Quantum dots are nanoscale semiconductor particles that exhibit quantum mechanical properties, such as discrete energy levels and quantized excitations. Arrays of quantum dots can be engineered to simulate strings with specific properties predicted by string theory. These arrays can be used to study and measure entanglement between quantum dots, providing insights into the behavior of entangled states in a controlled environment.

Importance in Quantum Entanglement

Quantum dot arrays offer a versatile platform for simulating and studying quantum entanglement. By precisely controlling the properties of individual quantum dots and their interactions, researchers can create and manipulate entangled states. These systems can be used to test theoretical predictions, explore new quantum phenomena, and develop practical applications in quantum computing and communication.

Experimental Setup

Quantum Dot Fabrication:
- Techniques: Quantum dots can be fabricated using various techniques such as molecular beam epitaxy (MBE), chemical vapor deposition (CVD), and colloidal synthesis. These methods allow for precise control over the size, shape, and

composition of the quantum dots, which in turn determine their quantum properties.
- Array Configuration: Quantum dots can be arranged in one-dimensional, two-dimensional, or three-dimensional arrays. The spacing and arrangement of the dots can be controlled to tune their interactions and simulate specific properties of strings.

Measurement and Control:
- Electrical and Optical Control: Quantum dots can be controlled using electrical gates and optical fields. Electrical gates can tune the energy levels of quantum dots, while optical fields can excite and manipulate their quantum states.
- Coupling and Interaction: The coupling between quantum dots can be controlled by adjusting their spacing and the strength of the applied fields. This coupling can be described by a Hamiltonian that includes terms for the interaction between neighboring dots:

$$H_{\text{int}} = \sum_{i,j} J_{ij} \left(\hat{a}_i^\dagger \hat{a}_j + \hat{a}_j^\dagger \hat{a}_i \right)$$

[86.0]

where J_{ij} is the coupling strength between quantum dots i is j, $\hat{a}_i^\dagger$ and $\hat{a}_i$ where the creation and annihilation operators for the quantum states of the dots are.

Mathematical Framework

Quantum State of Quantum Dot Arrays:-888888

- The quantum state of an array of N quantum dots can be represented by a many-body wavefunction $\Psi(x_1, x_2, \ldots, x_N)$, where x_i denotes the state of the i-th quantum dot.
- The Hamiltonian for the system can include terms for the individual quantum dots and their interactions:

$$H = \sum_i \left(\epsilon_i \hat{a}_i^\dagger \hat{a}_i\right) + \sum_{i,j} J_{ij} \left(\hat{a}_i^\dagger \hat{a}_j + \hat{a}_j^\dagger \hat{a}_i\right)$$

[87.0]

where ϵ_i is the energy level of the i-th quantum dot.

Entanglement Measures:

- Concurrence: For a pair of entangled quantum dots, the concurrence C can be used to quantify the degree of entanglement,

$$C = \max(0, \lambda_1 - \lambda_2 - \lambda_3 - \lambda_4)$$

[87.1]

where λ_i are the square roots of the eigenvalues of the matrix $\rho(\sigma_y \otimes \sigma_y)\rho^*(\sigma_y \otimes \sigma_y)$, and ρ is the density matrix of the two-qubit system.

- Entanglement Entropy: For larger arrays, the entanglement entropy S can be used:

$$S = -\text{Tr}(\rho \log \rho)$$

[87.2]

where ρ is the reduced density matrix of a subsystem.

Innovative Approaches

1. Hybrid Quantum Dot Systems:
    - Combine quantum dots with other quantum systems, such as superconducting qubits or trapped ions, to create hybrid systems with enhanced capabilities. These hybrid systems can leverage the strengths of different quantum technologies to study entanglement and other quantum phenomena.
    - Case: Integrate quantum dots with superconducting qubits to create a hybrid quantum processor that can perform complex quantum operations with high fidelity.

2. Machine Learning for Quantum Dot Control:
    - Optimize the control parameters for quantum dot arrays using machine learning algorithms. These algorithms can learn the optimal settings for creating and manipulating entangled states, improving the efficiency and accuracy of experiments.

Mathematical Framework:
    - Train a machine learning model to predict the optimal control parameters θ for achieving a desired entangled state:

$$\theta^* = \arg\max_{\theta} f(\rho(\theta))$$

[88.0]

where f is a function that quantifies the degree of entanglements, such as concurrence or entanglement entropy, and ρ(θ) is the density matrix as a function of the control parameters.

3. Dynamic Reconfiguration of Quantum Dot Arrays:
- Develop techniques for dynamically reconfiguring quantum dot arrays during experiments. This capability allows for the exploration of different interaction regimes and the study of the effects of varying coupling strengths and configurations on entanglement.
- Case: Use optical tweezers or electrostatic gates to move and reconfigure quantum dots in real-time, enabling the study of dynamic entanglement and decoherence processes.

Quantum dot arrays, a potent platform for simulating and studying quantum entanglement, are being explored using a range of innovative methods. These include advanced fabrication techniques, precise control methods, and novel approaches such as hybrid systems, machine learning, and dynamic reconfiguration. This exciting blend of techniques allows researchers to navigate the complex landscape of quantum phenomena as predicted by string theory and other quantum theories. The potential of these studies to enhance our understanding of entanglement and pave the way for practical quantum technologies is truly captivating.

A4. Superconducting Qubits:

Concept

Superconducting qubits are quantum circuits that leverage the properties of superconductors to create and manipulate quantum states. These qubits are particularly useful for studying quantum entanglement and other quantum phenomena because they can be engineered with specific properties and easily integrated into complex quantum systems. By using superconducting qubits with tailored characteristics, researchers can simulate string theory predictions and study the effects of these characteristics on entanglement.

Importance in Quantum Entanglement

Superconducting qubits provide a scalable and controllable platform for exploring quantum entanglement. Their engineered properties can mimic the behaviors predicted by string theory, such as specific vibrational modes and coupling constants. This enables detailed investigations into how these properties affect entanglement and coherence in quantum systems.

Experimental Setup

Fabrication and Design:
- Techniques: Superconducting qubits are typically fabricated using lithography techniques to pattern superconducting materials (like aluminum or niobium) on a silicon or sapphire substrate.
- Types of Qubits: Common designs include the Transmon, Fluxonium, and Phase qubits, each with unique characteristics suitable for different experiments.

Control and Measurement:
- Microwave Control: Superconducting qubits are controlled using microwave pulses that manipulate the quantum states by adjusting the frequency, phase, and amplitude of the microwave signals.
- Readout Mechanisms: The state of a qubit can be measured using dispersive readout techniques. The qubit state shifts the frequency of a coupled resonator, providing a detectable signal.

Interactions and Coupling:
- Capacitive and Inductive Coupling: Superconducting qubits can be coupled capacitively or inductively to control their interactions. The coupling strength g between two qubits can be adjusted to explore different interaction regimes.

Mathematical Representation:

$$H_{\text{int}} = \sum_{i,j} g_{ij} (\hat{\sigma}_i^+ \hat{\sigma}_j^- + \hat{\sigma}_i^- \hat{\sigma}_j^+)$$

[89.0]

where $\hat{\sigma}_i^+$ and $\hat{\sigma}_i^-$ are the raising and lowering operators for the i-th qubit, and g_{ij} is the coupling strength between qubits i and j.

Mathematical Framework

Qubit Hamiltonian:

- The Hamiltonian of a single superconducting qubit can be described by:

$$H_q = 4E_C(\hat{n} - n_g)^2 - E_J \cos(\hat{\phi})$$

[90.0]

where E_C is the charging energy, $\hat{n}$ is the number operator, n_g is the gate charge, E_J is the Josephson energy, and $\hat{\psi}$ is the phase operator.

Two-Qubit System:

- For a system of two coupled qubits, the Hamiltonian includes terms for both qubits and their interaction:

$$H = H_{q1} + H_{q2} + H_{\text{int}}$$

[90.1]

where H_{q1} and H_{q2} are the Hamiltonians of the individual qubits, and H_{int} describes their interaction.

Entanglement Measures:
- Concurrence: The concurrence C for a two-qubit system is used to quantify entanglement,

$$C = \max(0, \lambda_1 - \lambda_2 - \lambda_3 - \lambda_4)$$

[91.0]

where λ_i are the square roots of the eigenvalues of the matrix $\rho(\sigma_y \otimes \sigma_y)\rho^*(\sigma_y \otimes \sigma_y)$, and ρ is the density matrix of the system.

Innovative Approaches

1. Engineered Coupling Constants:
   - Design superconducting qubits with adjustable coupling constants to simulate the varying interaction strengths predicted by string theory. This can be achieved using tunable couplers that allow real-time adjustment of the coupling strength.
   - Case: A flux-tunable coupler can vary the inductive coupling between qubits, enabling the study of how different coupling constants affect entanglement.

2. Multi-Qubit Entanglement:
   - Extend the study to multi-qubit systems to explore more complex entangled states. Arrays of superconducting qubits can be used to create GHZ states, W states, and other multi-qubit entangled states.

Mathematical Representation:

$$|\text{GHZ}\rangle = \frac{1}{\sqrt{2}}(|000\rangle + |111\rangle)$$

[92.0]

$$|W\rangle = \frac{1}{\sqrt{3}}(|001\rangle + |010\rangle + |100\rangle)$$

[92.1]

- Case: Create and measure GHZ states using an array of three or more superconducting qubits and analyze the robustness of these states under different coupling conditions.

3. Noise-Resilient Entanglement:
   - Investigate methods to enhance the resilience of entangled states to noise and decoherence. Techniques such as dynamical decoupling, error correction codes, and robust control protocols can be employed to protect entanglement.

Mathematical Framework:
   - Dynamical Decoupling: Periodic application of control pulses to refocus the qubit states and cancel out the effects of environmental noise:

$$H_{\text{eff}} = H - \sum_k \left(\frac{\pi}{\Delta t_k}\right) \sigma_x^k$$

[93.0]

   - Error Correction: Implementing quantum error correction codes to detect and correct errors without disturbing the entangled state,

$$|\psi_L\rangle = \alpha|0_L\rangle + \beta|1_L\rangle$$

[93.1]

where $|\psi_L\rangle$ is the logical qubit state encoded in multiple physical qubits.

Superconducting qubits provide a versatile and powerful platform for studying the effects of string characteristics on entanglement. By engineering specific properties, controlling interactions, and employing innovative approaches such as multi-qubit entanglement and noise resilience techniques, researchers can explore the rich landscape of quantum phenomena predicted by string theory and other quantum theories. These studies will deepen our understanding of entanglement and support the development of practical quantum technologies.

A5. High-Energy Physics Experiments:

Concept

High-energy physics experiments are essential for probing the fundamental nature of particles and their interactions. By conducting experiments in high-energy physics labs, researchers can create and analyze particles with controlled string properties. These experiments involve colliding particles at extremely high energies to produce new particles and states of matter that can be studied to test predictions from string theory and other advanced quantum theories.

Importance in Quantum Entanglement

High-energy physics experiments provide a unique environment to explore quantum entanglement at energy scales and conditions that are not accessible in typical laboratory settings. By generating particles with specific string properties, researchers can study how these properties influence entanglement and other quantum phenomena. These experiments can validate theoretical models and uncover new aspects of quantum mechanics.

Experimental Setup

Particle Accelerators:
- Techniques: High-energy particle accelerators such as the Large Hadron Collider (LHC) at CERN, the Relativistic Heavy Ion Collider (RHIC), and the upcoming International Linear Collider (ILC) are designed to accelerate particles to nearly the speed of light and collide them with great precision.

- Collision Dynamics: In these collisions, the kinetic energy of the particles is converted into mass and energy of new particles, allowing the study of high-energy physics and quantum states predicted by string theory.

Control and Measurement:
- Detectors: Advanced detectors, such as the ATLAS and CMS detectors at the LHC, track and measure the properties of particles produced in collisions. These detectors provide detailed information about the particles' energy, momentum, charge, and other properties.
- Data Analysis: High-performance computing and data analysis techniques are employed to process the vast amount of data generated in these experiments. Machine learning algorithms and statistical methods are often used to identify and analyze rare events and correlations.

Interactions and String Properties:
- String Characteristics: By carefully selecting the initial conditions and parameters of the collisions, researchers can generate particles that exhibit specific string properties, such as particular vibrational modes or coupling constants.

Mathematical Representation:

$$E_{\text{collision}} = \sqrt{s}$$

[94.0]

where $\sqrt{s}$ is the center-of-mass energy of the collision, determining the energy available to produce new particles.

Mathematical Framework

Cross-Section and Event Rates:

- The probability of producing a particular final state in a high-energy collision is given by the cross-section σ,

$$\sigma = \int |\mathcal{M}|^2 \, d\Pi$$

[94.1]

where $\mathcal{M}$ is the matrix element for the process and $d\Pi$ is the phase space factor for the final state particles.

- The event rate R is given by:

$$R = \sigma \cdot \mathcal{L}$$

[94.2]

where $\mathcal{L}$ is the luminosity of the collider.

String Theory Predictions:

- String theory predicts specific signatures in high-energy collisions, such as the production of Regge trajectories or Kaluza-Klein modes. The properties of these particles can be described by the string tension T and the compactification radius R:

$$E_n = \sqrt{n^2 + \left(\frac{mR}{\hbar c}\right)^2}$$

[95.0]

where E_n is the energy of the n-th vibrational mode.

Innovative Approaches

1. Advanced Detector Technologies:
   - Develop and implement new detector technologies that can provide higher resolution and sensitivity. Cases include next-generation silicon detectors, time projection chambers, and calorimeters with improved granularity.
   - Case: Silicon photomultipliers (SiPMs) in calorimeters enhance photon detection efficiency and timing resolution, allowing for more precise measurements of particle energies and interactions.

2. Machine Learning for Data Analysis:
   - Apply machine learning algorithms to analyze collision data and identify patterns indicative of string properties. These algorithms can help detect rare events and improve the accuracy of particle identification.

Mathematical Framework:
   - Train machine learning models to classify collision events based on their features:

$$\hat{y} = f(\mathbf{x}; \theta)$$

[96.0]

where $\mathbf{x}$ represents the features of the event, $\hat{y}$ is the predicted classification, and θ are the parameters of the model.
   - Case: Use convolutional neural networks (CNNs) to analyze images from detectors and identify events corresponding to specific string theory predictions.

3. Quantum Simulation of High-Energy Processes:
- Use quantum computers to simulate high-energy particle collisions and predict the outcomes of experiments. Quantum simulations can provide insights into the behavior of quantum systems at high energies and help design experiments.

Mathematical Framework:
- The quantum simulation of a high-energy process can be represented by a Hamiltonian H that includes the interactions of interest:

$$H = H_{\text{kin}} + H_{\text{int}}$$

[97.0]

where H_{kin} describes the kinetic energy and H_{int} describes the interactions.

- Case: Simulate the scattering amplitudes of particles in string theory using a quantum computer to predict the signatures of new particles and guide experimental searches.

High-energy physics experiments, which are crucial for exploring the properties of particles predicted by string theory and other advanced quantum theories, are a testament to the invaluable role of researchers in advancing our understanding of the universe. By conducting experiments in high-energy physics labs, researchers can create and analyze particles with controlled string properties, study their interactions, and measure their entanglement. Innovative approaches such as advanced detector technologies, machine learning for data analysis, and quantum simulations offer new opportunities to enhance these experiments and uncover fundamental insights into the nature of quantum entanglement.

Potential Outcomes and Implications

B1. Positive Results:

Concept:

Positive results in high-energy physics experiments that demonstrate significant correlations between particles with shared string characteristics would provide strong evidence supporting the hypothesis. Such outcomes would suggest that quantum entanglement can be generalized beyond traditional systems to include particles with specific string properties.

Implications for Quantum Entanglement

Positive experimental results would have profound implications for our understanding of quantum mechanics and string theory. They indicate that entanglement is a more universal phenomenon than previously thought, potentially extending to a broader range of particles and interactions governed by string theory principles.

Mathematical Framework

Correlation Functions:
- The correlation function $E(\hat{a}, \hat{b})$ measures the degree of correlation between the measurement outcomes of two particles,

$$E(\hat{a}, \hat{b}) = \langle \psi | \hat{a} \otimes \hat{b} | \psi \rangle$$

[98.0]

where $\hat{a}$ and $\hat{b}$ are measurement operators, and $|\psi\rangle$ is the entangled state.

Bell's Inequality:
- A key test for entanglement is the violation of Bell's inequality. For two particles, the Bell parameter S is given by:

$$S = |E(\hat{a}, \hat{b}) - E(\hat{a}, \hat{b}') + E(\hat{a}', \hat{b}) + E(\hat{a}', \hat{b}')|$$

[99.0]

If $S > 2$, it indicates entanglement.

String Theory Characteristics:
- Particles with string characteristics, such as specific vibrational modes or compactification effects, can exhibit unique correlation patterns. The energy levels E_n of these modes can be described by,

$$E_n = \sqrt{n^2 + \left(\frac{mR}{\hbar c}\right)^2}$$

[100.0]

where n is the mode number, m is the mass, R is the compactification radius, $\hbar$ is the reduced Planck constant and c is the speed of light.

Potential Positive Outcomes

1. Observation of High Correlation:
- Experiments might reveal strong correlations between particles that share specific string characteristics, such as identical vibrational modes or coupling constants. This would provide direct evidence that these properties contribute to entanglement.

- Case: Measuring the correlation function $E(\hat{a}, \hat{b})$ for particles produced in high-energy collisions and finding $S > 2$.

2. Verification of String Theory Predictions:
    - Positive results could confirm predictions made by string theory, such as the existence of Regge trajectories or Kaluza-Klein modes, which would support the broader applicability of quantum entanglement.
    - Case: Detecting particles with energy levels described by the string theory equation $E_n = \sqrt{n^2 + \left(\dfrac{mR}{\hbar c}\right)^2}$.

[101.0]

3. Enhanced Quantum Technologies:
    - Demonstrating that string characteristics can lead to entanglement might enable new quantum technologies, such as advanced quantum sensors or communication systems that leverage these properties.
    - Case: Developing quantum sensors that utilize entangled particles with specific string properties for higher precision measurements.

Innovative Approaches

1. Real-Time Data Analysis:
    - Implement real-time data analysis techniques using machine learning to monitor and identify correlations continuously during high-energy experiments. This approach can quickly validate the presence of entanglement and adjust experimental parameters dynamically.

Mathematical Framework:

- Use machine learning models to predict the Bell parameter S from real-time data,

$$\hat{S} = f(\mathbf{x}; \theta)$$

[102.0]

where x represents the measured data and θ are the model parameters.

- Case: Deploy neural networks that analyze collision data on the fly to detect entanglement signatures.

2. Quantum Simulations for Prediction:
   - Use quantum simulations to predict the outcomes of high-energy experiments. These simulations can model the expected correlations and entanglement properties of particles with string characteristics, guiding the design of experiments.

Mathematical Framework:

Simulate the Hamiltonian of the system,

$$H = H_{\text{kin}} + H_{\text{int}}$$

[103.0]

where H_{kin} describes the kinetic energy and H_{int} describes the interactions.

- Case: Simulate the scattering processes of particles in a high-energy collider to predict the correlation functions and identify optimal experimental conditions.

3. Cross-Disciplinary Collaborations:
   - Foster collaborations between high-energy physicists, string theorists, and quantum information scientists to develop integrated experimental and theoretical approaches. These collaborations can lead to innovative experimental designs and new insights into the nature of entanglement.
   - Case: Joint research projects that combine expertise in string theory, quantum computing, and experimental physics to explore new frontiers in quantum entanglement research.

Positive results from high-energy physics experiments that demonstrate significant correlations between particles with shared string characteristics not only confirm our existing knowledge but also open up new avenues for exploration. These outcomes have the potential to revolutionize our understanding of quantum mechanics and string theory, leading to new quantum technologies and deeper insights into the fundamental nature of the universe.

B2. Negative Results:

Concept:

Negative results in high-energy physics experiments, where no significant correlations are observed between particles with shared string characteristics, indicate that the current hypothesis may need refinement. Such outcomes suggest that additional factors influencing entanglement need to be considered, and the experimental and theoretical models may require modification.

Implications for Quantum Entanglement

A lack of significant correlations would highlight the complexity of quantum entanglement and suggest that our understanding of the conditions required for entanglement is incomplete. It would prompt further investigation into additional variables and mechanisms that could influence entanglement beyond the currently hypothesized string characteristics.

Mathematical Framework

Hypothesis Testing:

- The null hypothesis H_0 in the context of these experiments is that there are no significant correlations between particles with shared string characteristics. The alternative hypothesis H_1 is that such correlations do exist.
- Statistical significance is typically assessed using a p-value. If the p-value is above a certain threshold (e.g., 0.05), H_0 is not rejected, indicating that the results are not statistically significant.

Correlation Functions and Bell's Inequality:

- If the correlation function $E(\hat{a}, \hat{b})$ does not violate Bell's inequality,

$$S \leq 2$$

[104.0]

where S is the Bell parameter it indicates that the observed correlations can be explained by classical physics, not quantum entanglement.

Potential Negative Outcomes

1. Lack of Significant Correlation:
   - If experiments do not reveal significant correlations between particles with shared string characteristics, it may suggest that these characteristics alone are not sufficient to induce entanglement.
   - Case: Measuring the correlation function $E(\hat{a}, \hat{b})$ for particles and finding $S \leq 2$, implying no entanglement.

2. Mismatch with String Theory Predictions:
   - If particles with expected string properties (like specific vibrational modes) do not exhibit the predicted entanglement behaviors, it may indicate a need to revisit the theoretical assumptions.
   - Case: Failing to detect energy levels described by $E_n = \sqrt{n^2 + \left(\frac{mR}{\hbar c}\right)^2}$ in particles that should exhibit these properties.

3. Inconclusive Results:
   - Results that are statistically inconclusive (high p-value) may indicate that the experimental setup or measurement techniques are insufficiently sensitive or that other unaccounted factors are at play.
   - Case: Experiments yielding p-values greater than 0.05, suggesting no significant deviation from the null hypothesis.

Innovative Approaches to Address Negative Results

1. Refining Experimental Techniques:
   - Improve the sensitivity and precision of experimental setups. This could involve upgrading detectors, using more sophisticated data analysis techniques, and ensuring better control over experimental variables.
   - Case: Enhancing detector resolution to reduce noise and improve the accuracy of correlation measurements.

2. Exploring Additional Variables:
   - Investigate other factors that could influence entanglement, such as environmental interactions, decoherence effects, or additional quantum properties not previously considered.

Mathematical Framework:
   - Consider a more comprehensive model of the system that includes environmental interactions

$$H_{\text{total}} = H_{\text{system}} + H_{\text{environment}} + H_{\text{interaction}}$$

[105.0]

where $H_{interaction}$ accounts for the coupling between the system and its environment.

   - Case: Studying the impact of temperature, electromagnetic fields, or material properties on the entanglement of particles.

3. Theoretical Revisions and Extensions:
- Re-evaluate and extend theoretical models to include more complex or previously overlooked aspects of string theory or quantum mechanics. This could involve higher-order corrections, alternative compactification schemes, or new interaction terms.

Mathematical Framework:
- Extend the string theory model to include additional dimensions or modified interactions

$$E_n = \sqrt{n^2 + \left(\frac{mR}{\hbar c}\right)^2 + \sum_k \lambda_k \phi_k}$$

[106.0]

where λ_k and ϕ_k represent new interaction terms.
- Case: Proposing new types of particles or vibrational modes that could account for the lack of observed entanglement.

Interdisciplinary Approaches

Concept

Exploring insights from other fields, such as condensed matter physics, quantum information science, and materials science, can provide new perspectives and deepen our understanding of quantum entanglement and string theory predictions. These interdisciplinary approaches can lead to the development of novel quantum algorithms for analyzing experimental data and can help explore similar phenomena in different physical systems.

Importance of Interdisciplinary Approaches

By integrating knowledge and techniques from various scientific disciplines, researchers can develop more robust hypotheses and experimental designs. This interdisciplinary approach can uncover previously overlooked factors influencing entanglement and enable the discovery of new quantum phenomena.

Contributions from Different Fields

1. Condensed Matter Physics:
   - Concept: Condensed matter physics studies the properties of matter in condensed phases, such as solids and liquids. It explores phenomena like superconductivity, magnetism, and quantum Hall effects, which are relevant to understanding entanglement in complex systems.
   - Case: The study of topological insulators in condensed matter physics has revealed robust edge states that are protected by topological order, providing insights into how topological properties can influence entanglement.

Mathematical Framework:

- The Hamiltonian for a topological insulator can be expressed as

$$H = \sum_{\mathbf{k}} \Psi_{\mathbf{k}}^{\dagger} [\mathbf{d}(\mathbf{k}) \cdot \boldsymbol{\sigma}] \Psi_{\mathbf{k}}$$

[107.0]

where $\Psi_{\mathbf{k}}$ is the electron field, $\mathbf{d}(\mathbf{k})$ is a vector function of momentum $\mathbf{k}$, and σ are the Pauli matrices.

2. Quantum Information Science:
   - Concept: Quantum information science focuses on the storage, manipulation, and transmission of information using quantum systems. It provides tools for analyzing and optimizing quantum entanglement and developing quantum algorithms.
   - Case: Quantum error correction codes are crucial for maintaining coherence in quantum computers and can be applied to protect entangled states in high-energy physics experiments.

Mathematical Framework:
- A common quantum error correction code is the stabilizer code, defined by a set of stabilizer generators S_i,

$$S_i |\psi\rangle = |\psi\rangle \quad \forall i$$

[108.0]

where S_i are Pauli operators that stabilize the code space.

3. Materials Science:
   - Concept: Materials science explores the properties and applications of materials, including their quantum mechanical behaviors. Understanding the material properties at the quantum level can enhance the design of experimental setups and detection technologies.
   - Case: The development of new materials with tailored electronic, optical, or magnetic properties can improve the sensitivity and precision of quantum sensors used in high-energy physics experiments.

Mathematical Framework:
   - The electronic properties of materials can be described using the tight-binding model:

$$H = -t \sum_{\langle i,j \rangle} (c_i^\dagger c_j + \text{H.c.}) + \sum_i \epsilon_i c_i^\dagger c_i$$

[109.0]

where t is the hopping parameter, ϵ_i is the on-site energy, and $c_i^\dagger$ and c_i are the creation and annihilation operators, respectively.

Development of Novel Quantum Algorithms

1. Machine Learning for Data Analysis:
   - Develop machine learning algorithms to analyze the vast amount of data generated in high-energy physics experiments. These algorithms can identify patterns and correlations indicative of quantum entanglement.

Mathematical Framework:
   - Machine learning models, such as neural networks, can be trained to recognize entanglement patterns

$$\hat{y} = f(\mathbf{x}; \theta)$$

[110.0]

where x represents input data features, $\hat{y}$ is the output prediction (*e.g., presence of entanglement*), and θ are the model parameters.

2. Quantum Simulation Algorithms:
   - Use quantum algorithms to simulate high-energy processes and predict experimental outcomes. These simulations can help design experiments and interpret results.

Mathematical Framework:
   - Quantum simulations can solve the Schrödinger equation for complex systems:

$$i\hbar \frac{\partial}{\partial t} |\psi(t)\rangle = H |\psi(t)\rangle$$

[111.00]

where H is the Hamiltonian of the system, and $|\psi(t)\rangle$ is the state vector.

3. Optimization of Quantum Experiments:
- Develop quantum algorithms for optimizing experimental parameters to maximize the detection of entanglement. These algorithms can find the best settings for variables such as collision energy, particle types, and detector configurations.

Mathematical Framework:
- Use optimization algorithms such as gradient descent to minimize or maximize a cost function $J(\theta)$:

$$\theta_{t+1} = \theta_t - \eta \nabla_\theta J(\theta_t)$$

[112.0]

where η is the learning rate.

Innovative Approaches

1. Interdisciplinary Workshops and Collaborations:
- Organize workshops and collaborative projects that bring together experts from condensed matter physics, quantum information science, and materials science to share insights and develop integrated research strategies.

- Case: A collaborative research project involving physicists, materials scientists, and quantum information theorists to design a new generation of quantum sensors optimized for high-energy physics experiments.

2. Hybrid Quantum-Classical Simulations:
   - Implement hybrid quantum-classical simulation techniques that combine classical computing power with quantum algorithms to model complex systems more efficiently.

Mathematical Framework:
   - The hybrid approach can use classical preprocessing to reduce the problem size before applying quantum algorithms.

$$H_{\text{eff}} = H_{\text{classical}} + H_{\text{quantum}}$$

[113.0]

   - Case: Use classical simulations to approximate the initial conditions of a high-energy collision, followed by quantum simulations to model the quantum entanglement dynamics.

3. Advanced Material Development for Quantum Detectors:
   - Collaborate with materials scientists to develop advanced materials that enhance the performance of quantum detectors. These materials can improve the resolution, sensitivity, and efficiency of detecting entangled particles.
   - Case: Development of superconducting materials with higher critical temperatures and lower noise characteristics for use in superconducting qubits and quantum sensors.

By considering insights and techniques from condensed matter physics, quantum information science, and materials science, researchers can refine their hypotheses and develop more effective experimental approaches for studying quantum entanglement. The development of novel quantum algorithms, a promising avenue for data analysis, and the exploration of similar phenomena in different physical systems will not only enhance our understanding of entanglement but also support the advancement of quantum technologies, sparking new avenues of research.

One key strategy for gaining a deeper understanding of the complex nature of entanglement is through cross-disciplinary collaborations. Negative results in high-energy physics experiments, where no significant correlations are observed, are not setbacks but valuable starting points for such collaborations. By improving our experimental techniques, investigating additional variables, and revising theoretical models, we can pave the way for fruitful collaborations that can identify new factors influencing quantum entanglement and open up new directions for future research.

References

[1] Green, M. B., Schwarz, J. H., & Witten, E. (1987). *Superstring Theory* (Vol. 1 & 2). Cambridge University Press.

[2] Zwiebach, B. (2004). *A First Course in String Theory*. Cambridge University Press.

[3] Bell, J. S. (1964). On the Einstein Podolsky Rosen paradox. *Physics Physique Физика*, 1(3), 195-200.

[4] Nielsen, M. A., & Chuang, I. L. (2000). *Quantum Computation and Quantum Information*. Cambridge University Press.

[5] Scully, M. O., & Zubairy, M. S. (1997). *Quantum Optics*. Cambridge University Press.

[6] Hameroff, S. R., & Penrose, R. (2014). Consciousness in the Universe: A Review of the 'Orch OR' Theory. *Physics of Life Reviews*, 11(1), 39-78.

[7] Aspect, A., Dalibard, J., & Roger, G. (1982). Experimental Test of Bell's Inequalities Using Time-Varying Analyzers. *Physical Review Letters*, 49(25), 1804-1807.

[8] Hensen, B., et al. (2015). Loophole-free Bell inequality violation using electron spins separated by 1.3 kilometres. *Nature*, 526(7575), 682-686.

[9] Clauser, J. F., Horne, M. A., Shimony, A., & Holt, R. A. (1969). Proposed experiment to test local hidden-variable theories. *Physical Review Letters*, 23(15), 880-884.

[10] Schlosshauer, M. (2005). Decoherence, the measurement problem, and interpretations of quantum mechanics. *Reviews of Modern Physics*, 76(4), 1267-1305.

[11] Bohr, N. (1935). Can Quantum-Mechanical Description of Physical Reality be Considered Complete? *Physical Review*, 48(8), 696-702.

[12] Everett, H. (1957). "Relative State" Formulation of Quantum Mechanics. *Reviews of Modern Physics*, 29(3), 454-462.

[13] Ghirardi, G. C., Rimini, A., & Weber, T. (1986). Unified Dynamics for Microscopic and Macroscopic Systems. *Physical Review D*, 34(2), 470-491.

[14] Rovelli, C. (1996). Relational Quantum Mechanics. *International Journal of Theoretical Physics*, 35(8), 1637-1678.

[15] Bohm, D. (1952). A Suggested Interpretation of the Quantum Theory in Terms of "Hidden" Variables. I and II. *Physical Review*, 85(2), 166-193.

[16] Zwiebach, B. (2004). *A First Course in String Theory*. Cambridge University Press.

[17] Polchinski, J. (1998). *String Theory* (Vol. 1 & 2). Cambridge University Press.

[18] Becker, K., Becker, M., & Schwarz, J. H. (2007). *String Theory and M-Theory: A Modern Introduction*. Cambridge University Press.

[19] Maldacena, J. (1998). The Large N Limit of Superconformal Field Theories and Supergravity. *Advances in Theoretical and Mathematical Physics*, 2, 231-252.

[20] Briegel, H.-J., & Zoller, P. (2000). Quantum Repeaters: The Role of Imperfect Local Operations in Quantum Communication. *Physical Review Letters*, 84(24), 5102-5105.

[21] Legero, T., Wilk, T., Kuhn, A., & Rempe, G. (2003). Time-resolved two-photon quantum interference. *Applied Physics B*, 77(8), 797-802.

[22] Wineland, D. J., et al. (1998). Experimental Issues in Coherent Quantum-State Manipulation of Trapped Atomic Ions. *Journal of Research of the National Institute of Standards and Technology*, 103(3), 259-328.

[23] The Intriguing Interplay of Electromagnetism and Weak Nuclear Force. https://www.trillmag.com/news/science/the-intriguing-interplay-of-electromagnetism-and-weak-nuclear-force/

[24] Witten, E. (1989). Quantum Field Theory and the Jones Polynomial. *Communications in Mathematical Physics*, 121(3), 351-399.

[25] Freed, D. S., & Moore, G. W. (2013). *The Uncertainty of Fluxes*. Communications in Mathematical Physics, 326(2), 459-476.

[26] Hasan, M. Z., & Kane, C. L. (2010). Colloquium: Topological Insulators. *Reviews of Modern Physics*, 82(4), 3045-3067.

[27] Gubser, S. S., & Karch, A. (2009). From gauge-string duality to strong interactions: A Pedestrian's Guide. *Annual Review of Nuclear and Particle Science*, 59, 145-168.

[28] Peskin, M. E., & Schroeder, D. V. (1995). *An Introduction to Quantum Field Theory*. Addison-Wesley.

[29] Weinberg, S. (1996). *The Quantum Theory of Fields* (Vol. 1-3). Cambridge University Press.

[30] Aad, G., et al. (2012). Observation of a new particle in the search for the Standard Model Higgs boson with the ATLAS detector at the LHC. *Physics Letters B*, 716(1), 1-29.

[31] Chatrchyan, S., et al. (2012). Observation of a new boson at a mass of 125 GeV with the CMS experiment at the LHC. *Physics Letters B*, 716(1), 30-61.

[32] Baur, U., Buice, M., & Berger, E. L. (2001). Multi-boson production at Hadron Colliders. *Physical Review D*, 64(9), 094022.

[33] Ladd, T. D., Jelezko, F., Laflamme, R., Nakamura, Y., Monroe, C., & O'Brien, J. L. (2010). Quantum computers. *Nature*, 464(7285), 45-53.

[34] Buchmueller, O., & Dolan, M. J. (2016). Searches for supersymmetry at the LHC in light of the discovery of the Higgs boson. *Annual Review of Nuclear and Particle Science*, 66, 271-300.

[35] Kelly, J., et al. (2015). State preservation by repetitive error detection in a superconducting quantum circuit. *Nature*, 519(7541), 66-69.

[36] Arute, F., et al. (2019). Quantum supremacy using a programmable superconducting processor. *Nature*, 574(7779), 505-510.

[37] Clarke, J., & Wilhelm, F. K. (2008). Superconducting quantum bits. *Nature*, 453(7198), 1031-1042.

[38] Devoret, M. H., & Schoelkopf, R. J. (2013). Superconducting circuits for quantum information: An outlook. *Science*, 339(6124), 1169-1174.

[39] Koch, J., Yu, T. M., Gambetta, J., Houck, A. A., Schuster, D. I., Majer, J., ... & Girvin, S. M. (2007). Charge-insensitive qubit design derived from the Cooper pair box. *Physical Review A*, 76(4), 042319.

[40] Gambetta, J. M., Córcoles, A. D., Merkel, S. T., Johnson, B. R., Smolin, J. A., Chow, J. M., ... & Steffen, M. (2012). Characterization of addressability by simultaneous randomized benchmarking. *Physical Review Letters*, 109(24), 240504.

[41] Kelly, J., Barends, R., Fowler, A. G., Megrant, A., Jeffrey, E., White, T. C., ... & Cleland, A. N. (2015). State preservation by repetitive error detection in a superconducting quantum circuit. *Nature*, 519(7541), 66-69.

[42] Dine, M. (2007). *Supersymmetry and String Theory: Beyond the Standard Model*. Cambridge University Press.

[43] Malka, V. (2002). Principles and applications of compact laser–plasma accelerators. *Nature Physics*, 4(6), 447-453.

[44] Reiserer, A., Kalb, N., Rempe, G., & Ritter, S. (2014). A quantum gate between a flying optical photon and a single trapped atom. *Nature*, 508(7495), 237-240.

[45] Loss, D., & DiVincenzo, D. P. (1998). Quantum computation with quantum dots. *Physical Review A*, 57(1), 120-126.

[46] Ladd, T. D., Jelezko, F., Laflamme, R., Nakamura, Y., Monroe, C., & O'Brien, J. L. (2010). Quantum computers. *Nature*, 464(7285), 45-53.

[47] Hanson, R., & Awschalom, D. D. (2008). Coherent manipulation of single spins in semiconductors. *Nature*, 453(7198), 1043-1049.

[48] Awschalom, D. D., Hanson, R., Wrachtrup, J., & Zhou, B. B. (2018). Quantum technologies with optically interfaced solid-state spins. *Nature Photonics*, 12(9), 516-527.

[49] Atatüre, M., Englund, D., Vamivakas, A. N., Lee, S. Y., & Wrachtrup, J. (2018). Material platforms for spin-based photonic quantum technologies. *Nature Reviews Materials*, 3(5), 38-51.

[50]. Childress, L., & Hanson, R. (2013). Diamond NV centers for quantum computing and quantum networks. *MRS Bulletin*, 38(2), 134-138.

[51] Paris, M. G. A., & Rehacek, J. (2004). *Quantum State Estimation*. Springer.

[52]. Ludlow, A. D., Boyd, M. M., Ye, J., Peik, E., & Schmidt, P. O. (2015). Optical atomic clocks. *Reviews of Modern Physics*, 87(2), 637-701.

[53] [Degen, C. L., Reinhard, F., & Cappellaro, P. (2017). Quantum sensing. *Reviews of Modern Physics*, 89(3), 035002.

[54] [Giovannetti, V., Lloyd, S., & Maccone, L. (2011). Advances in quantum metrology. *Nature Photonics*, 5(4), 222-229.

[55] Valamontes, Antonios. (2024a, May). Accessible Research Papers for Individuals with Dyslexia and Autism Spectrum Disorders. http://dx.doi.org/10.13140/RG.2.2.34352.37126

[56] Giovannetti, V., Lloyd, S., & Maccone, L. (2011). Advances in quantum metrology. *Nature Photonics*, 5(4), 222-229.

[57] Nielsen, M. A., & Chuang, I. L. (2000). *Quantum Computation and Quantum Information*. Cambridge University Press.

[58] Degen, C. L., Reinhard, F., & Cappellaro, P. (2017). Quantum sensing. *Reviews of Modern Physics*, 89(3), 035002.

[59] Preskill, J. (2018). *Quantum Computing in the NISQ era and beyond*. Quantum, 2, 79.

[60] Paris, M. G. A., & Rehacek, J. (2004). Quantum State Estimation. Spring

www.ingramcontent.com/pod-product-compliance
Lightning Source LLC
Chambersburg PA
CBHW062216220526
45471CB00009B/3230